SCIENCE AND HUMAN ORIGINS

SCIENCE AND HUMAN ORIGINS

ANN GAUGER

DOUGLAS AXE

CASEY LUSKIN

SEATTLE   DISCOVERY INSTITUTE PRESS   2012

Description

Evidence for a purely Darwinian account of human origins is supposed to be overwhelming. But is it? In this provocative book, three scientists challenge the claim that undirected natural selection is capable of building a human being, critically assess fossil and genetic evidence that human beings share a common ancestor with apes, and debunk recent claims that the human race could not have started from an original couple.

Copyright Notice

Publisher's Note

This book is part of a series published by the Center for Science & Culture at Discovery Institute in Seattle. Previous books include *The Deniable Darwin* by David Berlinski, *In the Beginning and Other Essays on Intelligent Design* by Granville Sewell, *Alfred Russel Wallace: A Rediscovered Life* by Michael Flannery, *The Myth of Junk DNA* by Jonathan Wells, and *Signature of Controversy*, edited by David Klinghoffer.

Library Cataloging Data

Science and Human Origins by Ann Gauger, Douglas Axe, and Casey Luskin

Illustrations by Jonathan Aaron Jones and others as noted.

124 pages

Library of Congress Control Number: 2012934836

BISAC: SCI027000 SCIENCE / Life Sciences / Evolution

BISAC: SCI029000 SCIENCE / Life Sciences / Genetics & Genomics

ISBN-13: 978-1-936599-04-2 (paperback)

Publisher Information

Discovery Institute Press, 208 Columbia Street, Seattle, WA 98104

Internet: http://www.discoveryinstitutepress.com/

Published in the United States of America on acid-free paper.

First Edition, First Printing: April 2012.

Cover Design: Brian Gage

Interior Layout: Michael W. Perry

CONTENTS

INTRODUCTION

G. K. CHESTERTON PUT IT WELL IN THE EVERLASTING MAN: "MAN is not merely an evolution but rather a revolution."[1]

Chesterton's comment neatly captures the unease many people have felt about Darwinian explanations of human origins right from the start. Even Alfred Russel Wallace, co-founder with Darwin of the theory of evolution by natural selection, eventually rejected a fully Darwinian explanation of human beings, preferring a form of intelligent design as an alternative.[2]

Since Darwin first proposed his theory of unguided evolution more than a century-and-a-half ago, similar doubts have been expressed by a parade of other scientists, philosophers, and public intellectuals.

Yet in recent years the public has been told—repeatedly—that the case for a purely Darwinian account of human origins is now beyond dispute. Indeed, hardly a month goes by without a new fossil fragment or scientific study being touted as further incontestable proof that the evidence for human evolution is well nigh overwhelming.

But is the evidence for a Darwinian account of human origins really so persuasive?

In this book, three scientists tackle that question. Their findings may surprise you. Ann Gauger is a developmental and molecular biologist with research experience at MIT, the University of Washington, and Harvard University. Douglas Axe is a molecular biologist who has held research scientist positions at Cambridge University, the Cambridge Medical Research Council Centre, and the Babraham Institute in Cambridge. Casey Luskin holds a graduate degree in earth sciences from the University of California at San Diego and has conducted geological research at the Scripps Institute for Oceanography. All three have published work in peer-reviewed science journals. All three have done "bench" science, not just science writing.

And all three think Darwin's theory is inadequate to account for both human origins and human uniqueness.

Before going on, it might be helpful to define what is being talked about when this book refers to "Darwinian" evolution. In public discussions today, evolution is a slippery term that can mean anything from generic change over time (an idea no one disputes) to an undirected historical process of "survival of the fittest" leading from one-celled organisms to man.

Strictly speaking, modern Darwinian theory (often called "neo-Darwinism") has two key planks: **common descent** and **natural selection acting on unplanned genetic variations.**

Common descent is the idea that all animals now living have descended from one or a few original ancestors through a process Darwin called "descent with modification." According to this idea, not only humans and apes share an ancestor, but so do humans, clams, and fungi.

Natural selection is the idea of "survival of the fittest." Modern Darwinian theory combines natural selection with the insights of modern genetics: Randomly occurring mutations and recombinations in genes produce unplanned variations among individual organisms in a population. Some of these variations will help organisms survive and reproduce more effectively. Over time, these beneficial variations will come to dominate a population of organisms, and over even more time, these beneficial variations will accumulate, resulting in entirely new biological features and organisms.

As Darwin himself made clear, natural selection is an unintelligent process that is blind to the future. It cannot select new features based on some future goal or potential benefit. As a result, evolution in a Darwinian sense is "the result of an unguided, unplanned process," to cite the words of 38 Nobel laureates who issued a statement defending Darwin's theory in 2005.[3]

In the Darwinian view, amazing biological features such as the vertebrate eye, or the wings of butterflies, or the blood-clotting system, are in no way the purposeful result of evolution. Rather, they are the unintended byproducts of the interplay of chance (random genetic mutations and recombinations) and necessity (natural selection). The same holds true for higher animals such as human beings. In the words of late Harvard paleontologist George Gaylord Simpson: "Man is the result of a purposeless and natural process that did not have him in mind."[4]

This book is focused on the *scientific* arguments about human evolution. But it should be obvious there is a larger cultural context to the debate.

Many secular Darwinians employ Darwin's theory as a battering ram to topple the idea of human exceptionalism. According to late Harvard paleontologist Stephen Jay Gould, Darwinian "biology took away our status as paragons created in the image of God."[5] Indeed, in the Darwinian view human beings are but "a fortuitous cosmic afterthought."[6] Princeton University bioethicist Peter Singer expresses a similar view. A champion of infanticide for handicapped human newborns, Singer makes clear that Darwinism supplies the foundation for his debased view of human beings: "All we are doing is catching up with Darwin. He showed in the 19th century that we are simply animals. Humans had imagined we were a separate part of Creation, that there was some magical line between Us and Them. Darwin's theory undermined the foundations of that entire Western way of thinking about the place of our species in the universe."[7] Darwin is likewise a patron saint for many radical environmentalists. In the approving words of former Earth First! activist Christopher Manes, "Darwin invited humanity to face the fact that the observation of nature has revealed not one scrap of evidence that humankind is superior or special, or even particularly more interesting than, say, lichen."[8]

Many religious Darwinists, meanwhile, use Darwinian science to urge revisions in traditional Christian teachings about both God and man. Karl Giberson, a co-founder of the pro-theistic-evolution BioLogos Foundation, argues that human beings were evil from the start because evolution is driven by selfishness; therefore, Christians must abandon the idea that human beings were originally created by God morally good.[9] Current BioLogos president Darrel Falk urges Christians to scrap their outdated belief in Adam and Eve as parents of the human race, claiming that evolutionary biology now proves "there was never a time when there was a single first couple, two people who were the progenitors of the entire human race."[10] And geneticist Francis Collins, the original inspiration for BioLogos, puts forward a watered-down view of God's sovereignty over the natural world. In one part of his book *The Language of God*, Collins claims (wrongly) that the human genome is riddled with functionless "junk DNA," which he claims is evidence against the idea that human beings were specifically designed by God.[11] Elsewhere in his book, Collins states that God "could" have known and specified the outcomes of evolution; but in that case, Collins believes that God made evolution *look* like "a random and undirected process," turning God into a cosmic trickster who creates the world by a process meant to mislead us.[12]

Biologist Kenneth Miller, author of *Finding Darwin's God*, goes considerably further. Miller explicitly argues that God neither knows nor directs the specific outcomes of evolution—including human beings. In Miller's view, "mankind's appearance on this planet was *not* preordained... we are here not as the products of an inevitable procession of evolutionary success, but as an afterthought, a minor detail, a happenstance in a history that might just as well have left us out."[13] According to Miller, God did know that undirected evolution would produce some sort of rational creature eventually, but the creature produced by evolution might have been a "a big-brained dinosaur" or "a mollusk with exceptional mental capabilities" rather than a human being.[14]

Whether secular or religious, these champions of modern Darwinian theory all share the same underlying assumption: In their view, science has proven Darwinian evolution beyond a shadow of a doubt;

therefore our understanding of human beings and the rest of life must be radically reshaped according to Darwinian tenets.

But what if this assumption turns out to be wrong? What if the unbounded faith placed in Darwinian theory—especially as applied to human beings—is scientifically unwarranted?

The authors of this volume invite you to consider that possibility.

- In chapters 1 and 2, Ann Gauger and Douglas Axe challenge the central claim that Darwin's undirected mechanism of natural selection is really capable of building a human being.

- In chapters 1, 3, and 4, Ann Gauger and Casey Luskin critically assess the genetic and fossil evidence that human beings share a common ancestor with apes.

- And in the final chapter, Ann Gauger refutes scientific claims that the human race could not have started from an original couple.

Although much of this book focuses on the shortcomings of Darwinian theory, the scientists represented here are not merely critics of the existing paradigm. Instead, they share a positive vision that much of biology would make better sense from the perspective of intelligent design rather than unguided Darwinian evolution. Often mischaracterized (and wrongly conflated with creationism), intelligent design is simply the effort to investigate empirically whether the exquisitely coordinated features we find throughout nature are the result of an intelligent cause rather than a blind and undirected process like natural selection.[15]

Because intelligent design focuses on whether the development of life was purposeful or blind, it directly challenges the second plank of Darwinian theory (unguided natural selection) rather than the first (common descent). Nevertheless, intelligent design scientists remain free to critically assess the actual evidence for common descent, as they do here.

Whether you consider yourself secular, religious, or something in between, the science of human origins raises deep and continuing questions about what it means to be human. You are invited to explore some of these questions in the pages that follow.

John G. West, Ph.D.
Associate Director, Center for Science and Culture
Discovery Institute, Seattle

ENDNOTES

1. G. K. Chesterton, *The Everlasting Man* (San Francisco: Ignatius Press, 1993), 26.

2. See Michael Flannery, *Alfred Russel Wallace: A Rediscovered Life* (Seattle: Discovery Institute Press, 2011).

3. Letter from Nobel Laureates to Kansas State Board of Education, Sept. 9, 2005. The letter was sent out under the auspices of the Elie Wiesel Foundation. A copy or the letter was posted at http://media.ljworld.com/pdf/2005/09/15/nobel_letter.pdf (accessed Aug. 8, 2006).

4. George Gaylord Simpson, *The Meaning of Evolution: A Study of the History of Life and of Its Significance for Man*, revised edition (New Haven: Yale University Press, 1967), 345.

5. Stephen J. Gould, *Ever Since Darwin: Reflections in Natural History* (New York: W. W. Norton and Company, 1977), 147.

6. Stephen J. Gould, *Dinosaur in a Haystack: Reflections in Natural History* (New York: Harmony Books, 1995), 327.

7. Quoted in Johann Hari, "Peter Singer: Some people are more equal than others," *The Independent*, July 1, 2004, http://www.independent.co.uk/news/people/profiles/peter-singer-some-people-are-more-equal-than-others-6166342.html (accessed on March 6, 2012).

8. Christopher Manes, *Green Rage: Radical Environmentalism and the Unmaking of Civilization* (Boston: Little, Brown, and Company, 1990), 142.

9. Karl Giberson, *Saving Darwin: How to Be a Christian and Believe in Evolution* (New York: HarperOne, 2008), 11–13. The book has a Foreword by Francis Collins. For a discussion of Giberson's view, see John G. West, "Nothing New Under the Sun" in Jay Richards, *God and Evolution: Protestants, Catholics, and Jews Explore Darwin's Challenge to Faith* (Seattle: Discovery Institute Press, 2010), 33–52.

10. Darrel Falk, "BioLogos and the June 2011 'Christianity Today' Editorial," June 6, 2011, http://biologos.org/blog/biologos-and-the-june-2011-christianity-today-editorial (accessed March 6, 2012).

11. Francis S. Collins, *The Language of God: A Scientist Presents Evidence for Belief* (New York: Free Press, 2006), 135–136. For a rebuttal of some of Collins's scientific arguments, see chapter four of this book by Casey Luskin. Also see Jonathan Wells, "Darwin of the Gaps," in Richards, *God and Evolution*, 117–128.

12. Collins, *The Language of God*, 205–206.

13. Kenneth R. Miller, *Finding Darwin's God: A Scientist's Search for Common Ground Between God and Evolution* (New York: HarperCollins, 1999), 272.

14. Miller, quoted in John G. West, *Darwin Day in America: How Our Politics and Culture Have Been Dehumanized in the Name of Science* (Wilmington, DE: ISI Books, 2007), 226.

15. For good introductions to intelligent design, see Guillermo Gonzalez and Jay Richards, *The Privileged Planet: How Our Place in the Cosmos is Designed for Discovery* (Washington DC: Regnery, 2004); Stephen C. Meyer, *Signature in the Cell: DNA and the Evidence for Intelligent Design* (New York: HarperOne, 2009), and William Dembski and Jonathan Wells, *The Design of Life* (Dallas: Foundation for Thought and Ethics, 2008).

1

SCIENCE AND HUMAN ORIGINS

Ann Gauger

Explaining human origins requires a new way of approaching
things. There is no strictly neo-Darwinian path from a chimp-like
ancestor to us, no matter how similar we appear to be.

LATELY THE STORY OF HUMAN ORIGINS HAS BECOME A SUBJECT OF
renewed controversy in the media. In 2011 both National Public Radio and *Christianity Today* ran high-profile stories featuring Christian scholars who claim not only that human beings evolved from ape-like ancestors, but who assert that science has refuted the traditional Christian belief in a first human couple, Adam and Eve.[1] Apparently, these scholars are convinced that the neo-Darwinian account of our origins has now eliminated any need for other explanations. Equally apparently, the media thought this story was newsworthy because the people talking were Christians, who presumably had no bias against religion, and at least some of whom were credible scientists.

When I first saw these stories, it struck me how uncritically all these people accepted the scientific arguments for human evolution. This is a mistake. Science is not an error-free enterprise, so arguments need to be carefully evaluated. This is especially the case when it comes to a highly charged issue like human evolution.

Most of the argument *for* our common ancestry with ape-like creatures is based on similarity—similarity in anatomy, and similarity in DNA sequence. Yet I know from my own experiments that similarity between two complex structures does not reliably indicate an evolutionary path between them.

Similarity by itself says nothing about what mechanisms are responsible for the apparent relatedness, especially when substantial genetic change is necessary. In fact, there is a surprising disregard among evolutionary biologists for the amount of genetic change that would be needed to actually accomplish the evolutionary transitions they propose, and the amount of time it would require. As I shall explain, these obstacles are a significant factor in human evolution and indicate we cannot have come from an ape-like ancestor by any unguided process.

WHAT IS THE EVIDENCE FOR COMMON ANCESTRY?

THE IDEA of our gradual evolution from ape-like ancestors goes all the way back to Darwin himself, although in his time no transitional fossils were known to exist. Since Darwin's time, paleoanthropologists have uncovered fossil remains that appear to be intermediate in form between great apes and humans. These fossils, together with more recent DNA sequence comparisons from living species, have led to a proposed tree of common descent for the great apes and humans (together referred to as hominids):

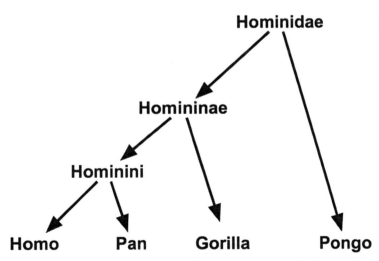

Figure 1-1: Currently accepted tree of common descent for hominids. *Illustration: Ann Gauger.*

The ends of the branches represent living genera (groups of species that share similar characteristics), and the branch points represent the last common ancestors from which the genera are thought to have come. The last branch on the tree, assigned the name Hominini, or hominins, includes *Pan* (chimpanzees), our hypothetical most recent common ancestor with *Pan*, all the supposed transitional species in between, and us.[2]

What evidence is there for this tree? As I said previously, it hinges on two kinds of data: **anatomical similarities and differences** among the great apes, fossil hominins and us; and **comparative analysis of DNA sequences** from living species. *It also depends on one very big but unproven assumption—that any similarities found are due to descent from a common ancestor.* It is that assumption I wish to challenge in this chapter.

The fossil evidence for our evolution from apes is actually quite sketchy.[3] Ancient hominin fossils are rare, and they typically consist of bone fragments or partial disarticulated skeletons obtained from different locations around the world and from different geologic strata. They fall into two basic categories: ape-like fossils, and *Homo*-like fossils. This discontinuity between fossil types is well-known. Nonetheless, the hominin fossils have been interpreted as historical, physical evidence of our common ancestry with apes. Ernst Mayr, a well-known evolutionary biologist, acknowledged both the gap and the story-telling in his book *What Makes Biology Unique*:

> The earliest fossils of *Homo*, *Homo rudolfensis* and *Homo erectus*, are separated from *Australopithecus* by a large, unbridged gap. How can we explain this seeming saltation? Not having any fossils that can serve as missing links, we have to fall back on the time-honored method of historical science, the construction of a historical narrative.[4]

The resulting historical narrative is familiar to us all, as depicted in drawings commonly found in *National Geographic* and similar magazines.

The evidence from DNA comparisons is similarly enigmatic. DNA sequences are strings of nucleotides millions or billions in length. Aligning DNA sequences in order to compare them is a tricky business. There can be single bases changes, insertions or deletions, duplications, and rearrangements of the DNA that complicate things and may or may not be included in comparisons.[5] The degree of similarity calculated depends on how the analysis is done, and what is excluded or included.[6] But putting aside arguments about *how* similar we are to chimps, the question is: What does similarity *demonstrate?*

For most biologists, similarity is *assumed* to confirm that humans and chimps are linked together by common ancestry. This assumption underlies all evolutionary reasoning. But note that similarity of structure or sequence cannot confirm common descent by itself. "Mustang" and "Taurus" cars have strong similarities, too, and you could argue that they evolved from a common ancestor, "Ford." But the similarities between these cars are the result of common design, not common ancestry.

For any story about common ancestry to be verified, including the proposed story of *our* common ancestry, two things must be shown. First, a step-wise adaptive path must exist from the ancestral form to the new form, whether it is to a new gene, a new protein, or a new species; and second, if it is to have happened by an unguided, neo-Darwinian mechanism, there must be enough time and probabilistic resources for neo-Darwinian processes to traverse that path. The neo-Darwinian mechanisms of mutation, recombination, genetic drift and natural selection must be enough to produce the proposed innovation in the time available. These two things, a step-wise, adaptive path, and enough time and probabilistic resources for the path to be traversed, are absolutely necessary for neo-Darwinian evolution to have occurred.

Yet these two things have yet to be demonstrated for any significant evolutionary transition. In what follows, I will show that these two things haven't been demonstrated for human evolution—and probably never will be.

An Experimental Test

How realistic is it for humans to have evolved by neo-Darwinian means? We can't go back and observe the past directly, so we need to assess the likelihood of much simpler transitions, the kinds of changes that are testable in the lab.

Proteins that look alike are commonly assumed to have a common evolutionary origin. If the proteins have different functions, then it is assumed that some sort of neo-Darwinian process led to their duplication and divergence. This is the story of common descent writ small. But unlike humans and chimps, proteins can be easily manipulated and tested in the lab for successful functional change. We can actually establish how many mutations are required to switch old proteins to new functions, and thus determine what kinds of innovations *are* possible according to the rules of neo-Darwinism. If the neo-Darwinian story fails here, it fails everywhere.

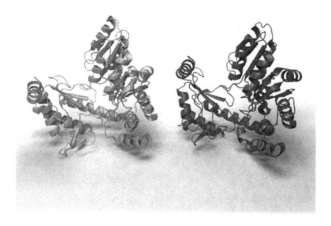

Figure 1-2: Kbl and BioF, two structurally similar proteins from *E. coli.*
Illustration: Ann Gauger and Douglas Axe.

My colleague Douglas Axe and I took two bacterial proteins that look a great deal alike, but have distinctly different functions. They are thought to be evolutionary cousins, descended from a common ancestor millions of years ago, *because of their similar structures.* These proteins, called Kbl and BioF, are shown in **Figure 1-2** above. Kbl and BioF are not directly descended from the other; nonetheless, a functional shift from something like Kbl to something like BioF must be possible if neo-Darwinism is true. Functional shifts like this one are found everywhere in families of related proteins, and so should be relatively easy to achieve.

Yet when we experimentally determined how many mutations it would take, we found that it would take *at least seven* mutations to evolve one enzyme into the other—too many mutations to have occurred by an unguided neo-Darwinian process.[7]

Bacteria are genetic workhorses for evolutionary research, precisely because they are capable of rapid adaptation—as long as it takes only one or two mutations. Three coordinated mutations are a stretch even for bacteria, if all of the intermediates are neutral (have no beneficial effect for the organism). But for one of our enzymes to evolve the other's function, it would take at least *seven* and probably many more mutations. The waiting time for seven coordinated neutral mutations to arise in a bacterial population is on the order of 10^{27} years. To put that in some sort of perspective, remember that the universe is only about 10^{10} years old.[8] It can't have happened.

Yet this is precisely the kind of transition that neo-Darwinism is meant to explain—structurally similar, yet functionally distinct proteins should be able to diverge by a process of mutation and selection. If this shift in function is not within reach of known neo-Darwinian mechanisms, something else must be going on.[9]

In case you are wondering, our result is in line with other published research on recruitment of proteins to new functions. Attempts to convert proteins to genuinely new functions typically require eight or more mutations, well beyond the reach of neo-Darwinian processes.

GETTING TO HUMAN

THE RESEARCH I described above has shown that similarity of structure is not enough to establish that there is an adaptive path between two proteins with distinct functions. In fact, it is likely to be the case that, in general, neo-Darwinian processes are not sufficient to produce genuine innovations, because too many specific mutations are required. Now we need to consider whether or not this analysis also applies to the necessary transitions to get from an ape-like ancestor to us.

Let's begin by considering what distinguishes us from great apes. What are our distinctive characteristics? There are significant anatomical differences, of course: Our upright walking, longer legs and shorter arms, changes in muscle strength, our significantly larger brains and skulls (three times bigger than great apes), and our refined musculature in hands, lips and tongues. There are also our relative hairlessness and changes to our eyes. More importantly, there are whole realms of intellect and experience that make us unique as humans. Abstract thought, art, music, and language: These things separate us from lower animals fundamentally, not just in degree but in kind.

How many mutations might be required to produce these kinds of innovations? We really have very little data by which to track intellectual changes, so let's consider just the physical characteristics that distinguish us from chimps.

Chimps are suited for life in the trees. Humans are suited for life on the ground, walking and running. The anatomical changes needed to move from tree-dwelling to complete terrestrial life are many. To walk and run effectively requires a new spine, a different shape and tilt to the pelvis, and legs that angle in from the hips, so we can keep our feet underneath us and avoid swaying from side to side as we move. We need knees, feet and toes designed for upright walking, and a skull that sits on top of the spine in a balanced position. (The dome of our skull is shifted rearward in order to accommodate our larger brain and yet remain balanced.) Our jaws and muscle attachments must be shifted, our face flattened, and the sinuses behind the face and the eye sockets located in different places, to permit a forward gaze and still be able to see where to put our feet.

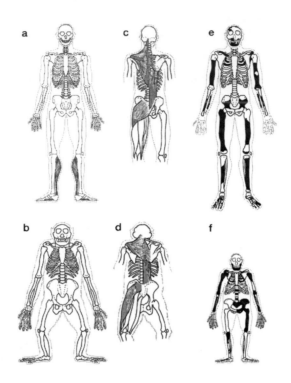

Figure 1-3: Comparison of hominin anatomy. Human [(a) and (c)] and chimp [(b) and (d)] skeletons and major muscles involved in running are shown. *H. erectus* (e) and *A. afarensis* (f) are drawn to the same scale, with the existing bones for each shown in black. White bones in (e) and (f) are hypothetical.

Illustration: Adapted by Jonathan Jones with permission from Macmillan Publishers Ltd: Nature, Dennis M. Bramble and Daniel E. Lieberman, "Endurance running and the evolution of Homo," 432 (2004): 345–352, Figure 3, copyright 2004.

Many of these differences can be seen in **Figure 1-3** above.[10] Humans (a,c), and chimps (b,d), have significantly different shoulders, rib cages, spines, pelvises, hips, legs, arms, hands and feet, each appropriate for different modes of living.

Now let's consider the proposed evolutionary story based on the fossil record. Shown next to the chimp and human figures are two hominin figures, *Homo erectus* (e), and *Austalopithicus afarensus* (f), reconstructed from the partial skeletal remains of "Turkana boy" from 1.6 million years ago (mya) and "Lucy" (3.2 mya), respectively.[11] In this drawing "Lucy" resembles a chimp in most respects, though her leg and pelvis bones suggest she spent time walking upright. Scientists debate the degree to which her lifestyle was terrestrial, however, as certain elements of her frame would have made walking inefficient.

"Turkana boy," in contrast, is much more similar in anatomy to modern humans. His hominin species, *Homo erectus*, first appeared in the fossil record about two mya ago, with all the adaptations in place for fully upright travel, including running over long distances. His only obvious difference from *Homo sapiens* is his skull, which though substantially larger than that of *A. afarensis*, is smaller than that of modern humans (though not outside the range of modern human genetic variation).

If our common ancestry with chimps is true, the transition to fully human must include something like the shift from *A. afarensis* to *H. erectus*. And here is where the discontinuity lies. *H. erectus* is the first fossil species with a nearly modern human anatomy and a constellation of traits not seen in any prior hominin. There simply is no good transitional species to bridge the gap. As John Hawks, a paleoanthropologist at the University of Wisconsin/Madison states:

> No australopithecine species is obviously transitional [to *Homo erectus*].... *Our interpretation is that the changes are sudden and interrelated* and reflect a bottleneck that was created because of the isolation of a small group from a parent australopithecine species. In this small population, a combination of drift and selection resulted in a *radical transformation* of allele frequencies, fundamentally *shifting the adaptive complex*; in other words, a *genetic revolution*.[12] [Emphasis added, internal citations removed for clarity.]

So Much to Do, So Little Time

FOR THE purposes of my argument, I don't intend to argue that *H. erectus* was or was not the first human being, or is directly part of our lineage. Instead, I want to focus on the anatomical changes that must be accomplished to go from *A. afarensis* to *H. erectus*. Regardless of whether or not other transitional hominins are found, these are the kinds of anatomical changes that must have occurred.

For a "radical transformation" of this kind to have happened by strictly neo-Darwinian means, as Hawks *et al.* imply, then some combination of mutation, genetic drift, and natural selection *must* be capable of producing the change. But "shifting the adaptive complex" to the new *H. erectus* anatomy would require reorganizing multiple anatomical structures, the kind of thing likely to require multiple specific mutations.

Two questions then arise: (1) How many mutations would it take to turn an australopithecine species into a *Homo erectus?* And (2) If there are only one and a half million years between *A. afarensis* and *H. erectus,* can neo-Darwinism produce the necessary changes in the time allotted?

How many mutations would it take? Bramble and Lieberman count sixteen features of the human body that first appear in *H. erectus* or *H. sapiens.*[13] These features are necessary to stabilize the head, permit counter-rotation of the torso with the head and hips, stabilize the trunk, absorb shock and transfer energy during running. Many of these changes must occur together to be of any benefit.

Is there enough time to get sixteen anatomical changes by a neo-Darwinian process? Each of these new features probably required multiple mutations. Getting a feature that requires six neutral mutations is the limit of what bacteria can produce. For primates (e.g., monkeys, apes and humans) the limit is much more severe. Because of much smaller effective population sizes (an estimated ten thousand for humans instead of a billion for bacteria) and longer generation times (fifteen to twenty years per generation for humans *vs.* a thousand generations per year for bacteria), it would take a *very* long time for even a *single* beneficial mutation to appear and become fixed in a human population.

You don't have to take my word for it. In 2007, Durrett and Schmidt estimated in the journal *Genetics* that for a single mutation to occur in a nucleotide-binding site[14] and be fixed in a primate lineage would require a waiting time of six million years.[15] The same authors later estimated it would take 216 million years for the binding site to acquire *two* mutations, if the first mutation was neutral in its effect.[16]

FACING FACTS

BUT SIX million years is the *entire time allotted* for the transition from our last common ancestor with chimps to us according to the standard evolutionary timescale. Two hundred and sixteen million years takes us back to the Triassic, when the very first mammals appeared. One or two mutations simply aren't sufficient to produce the necessary changes—sixteen anatomical features—in the time available. At most, a new binding site might affect the regulation of one or two genes. Durrett and Schmidt acknowledge the problem, and suggest that it can be overcome because there are an estimated 20,000 genes evolving independently, many of which might benefit from mutation(s) in their regulatory regions.

This is unreasonable. Having 20,000 genes available for change does not make the task easier. Many of the anatomical changes seen in *H. erectus* had to occur together to be of benefit. Individually they would be useless or even harmful. So even if a random mutation or two resulted in one change, they would be unlikely to be preserved. And getting all sixteen to appear and then become fixed within six million years, let alone the one and a half million that it apparently took, can't have happened through an unguided process.

To get an idea of why it won't work, imagine letting your toddler loose on your computer operating system, allowing her to randomly change 1s to 0s, or insert or delete stretches of 1s and 0s, or rearrange them in the code. How likely is it that she will develop a new subroutine that improves the function of the operating system? Unless you had the foresight to write an executive program that wipes out all changes that reduce the efficiency of the operating system or crash it, she *will* crash the system. Even with an executive program that eliminates crash-inducing changes and rewards efficiency, her haphazard changes are very unlikely to *ever* create a new subroutine. This is because the executive function has no foresight and can't see that certain changes, if preserved, may lead eventually to a valuable new subroutine.

That toddler is like mutation, and natural selection is like that executive function. Natural selection may be good at weeding out mistakes that make the system crash or reduce efficiency, but it's really bad at innovation. It has no foresight, and can't predict which changes could lead to an innovation and then preserve them. It lacks intention. In fact, natural selection often permits *the loss* of considerable functional genetic information if it gives some slight survival advantage in the current environment.[17]

Remember, any innovation that requires more than six specific neutral changes is impossible for bacteria, even with their rapid growth rates and large population sizes. For large mammals like us, the picture for neo-Darwinism is much, much bleaker.

How many mutations would it take to evolve the anatomical changes necessary for walking and running? Dozens if not hundreds or thousands—if it could happen by random mutation at all. If the time span available for human evolution from a chimp-like ancestor is six million years, the effective population size is ten thousand, the mutation rate is 10^{-8} per nucleotide per generation and the generation time is five to ten years (for a chimp-like ancestor), only a *single* change to a particular DNA binding site could be expected to arise. It strains credibility to think that *all sixteen* anatomical features evolved fortuitously in that same time frame, especially if each required multiple mutations. *Given these numbers, it is extremely improbable, if not absolutely impossible, for us to have evolved from hominin ancestors by a gradual, unguided process.*

HUMAN EXCEPTIONALISM

THE ABOVE argument was based solely on the anatomical changes required for fully upright, bipedal posture and efficient long-distance travel. But I cannot leave this discussion without pointing out the many other things that distinguish us from apes. At the fine motor level, we have many abilities that require anatomical features that apes lack —we have many more finely controlled muscles in our hands, face, and tongues, for example. Without them our dexterity as artists or craftsmen, our ability to converse, and our ability to express fine distinctions in emotion by our facial expressions would be impossible.

But even more significant are our cognitive and communicative abilities. We are much more than upright apes with fine motor control. Our capacity for abstract thought, self-conscious reflection, and ability to communicate put us in another category entirely. These attributes are orders of magnitude more complex than anything animals can do. For example, language requires both anatomical features (the position of our larynx and language centers in our brains), and a mysterious innate knowledge of the rules of grammar that appears to be hard-wired into our brain. Three-year-olds know these rules instinctively. Apes don't. True language requires the ability to think abstractly. Words are symbols that stand in for things and ideas. We communicate by arranging words into complex symbolic utterances. We think new thoughts and convey new ideas to others. We reflect on ourselves. We discuss our origins, write sonnets, and describe both imaginary worlds and the real one we inhabit. Language both reflects and enriches our capacity for abstract thinking and creativity.

Where did these massive increases in fine-motor dexterity, and the quantum leaps of language, art, and abstract thought come from? Our uniquely human attributes constitute a quantum leap, not just an innovation, a leap that cannot have arisen without guidance.[18] We are not souped-up apes.

Explaining our origin requires a new way of approaching things. There is no strictly neo-Darwinian path from a chimp-like ancestor to us, no matter how similar we appear to be. The mechanisms of random mutation, natural selection and genetic drift are insufficient to accomplish the needed changes in the time allotted, so other explanations need to be explored. Are we the product of some sort of necessary cosmic unfolding? The lucky result of an ever-ramifying series of universes? Or are we the embodiment of intelligent design by an agent or agents unknown?

When evaluating explanatory causes for beings such as ourselves, we need to choose a cause that is up to the task. I personally am convinced that unguided, unintelligent processes can't do the job, not only because the neo-Darwinian mechanism is utterly insufficient, but also because we are beings capable of intelligence and creativity. These qualities are what make us human, and together with our capacity for empathy and our desire for goodness and beauty, they point toward the kind of cause that *is* sufficient to explain our origins.

Figure 1-4 Humans are exceptional in their creativity, their artistry, and their exercise of reason.
Illustration: Annbale Caracci, "Studio di nudo maschile," public domain, reprinted from Wikimedia Commons.

ENDNOTES

1. See, for example, Barbara Bradley Hagerty, "Evangelicals Question the Existence of Adam and Eve," *National Public Radio,* August 9, 2011, accessed March 6, 2012, and Richard N. Ostling,"The Search for the Historical Adam," *Christianity Today,* June 2011, accessed March 6, 2012.

2. The tree was recently redrawn—and the terminology changed—to accommodate sequence data that (mostly) places us in our own group with chimps. Previously the same group was called the hominids, but that term now covers all great apes and us. Some articles still use the older terminology. See http://news.nationalgeographic.com/news/2001/12/1204_hominin_id.html.

3. For more details on the subject, see chapter 3 on "Human Origins and the Fossil Record" by Casey Luskin later in this volume.

4. Ernst Mayr, *What Makes Biology Unique?* (New York: Cambridge University Press, 2004), 198.

5. For a discussion of one kind of rearrangement that is often used as evidence for common descent, see chapter 4 by Casey Luskin on "Francis Collins, Junk DNA, and Chromosomal Fusion."

6. T. C. Wood, "The chimpanzee genome and the problem of biological similarity," *Occas Papers of the BSG* 7 (2006): 1–18; G. Glazko, et. al., "Eighty percent of proteins are different between humans and chimpanzees," *Gene 346* (2005): 215–219; J. Cohen, "Relative differences: The myth of 1%," *Science* 316 (2007): 1836.

7. A. K. Gauger and D. D. Axe, " The evolutionary accessibility of new enzyme functions: A case study from the biotin pathway," *BIO-Complexity* 2, no. 1 (2011): 1–17.

8. Ibid.

9. Douglas Axe amplifies this story to underscore the insufficiency of the neo-Darwinian engine to drive evolutionary change in the next chapter.

10. D. M. Bramble and D. E. Lieberman, "Endurance running and the evolution of *Homo,*" *Nature* 432 (2004): 345–352.

11. "Lucy" is 40% complete as a skeleton, with only a thigh bone and partial pelvis to reconstruct her lower limbs, while "Turkana boy" is missing only the hands and feet.

12. J. Hawks *et al.,* "Population bottlenecks and Pleistocene human evolution," *Mol Biol Evol* 17 (2000): 2–22.

13. Bramble and Lieberman, "Endurance running." For a list of hundreds of phenotypic traits in humans that differ from the great apes, see A. Varki and T. K. Altheide, "Comparing the human and chimpanzee genomes: Searching for needles in a haystack," *Genome Research* 15 (2005): 1746–1758.

14. A nucleotide-binding site is a piece of DNA eight nucleotides long. Durrett and Schmidt (see below) calculated how long it would take for a single mutation to generate a seven out of eight match for an eight nucleotide binding site (with six out of eight nucleotides already correct) in a stretch of DNA one

thousand nucleotides long. Creation of such a binding site might affect the behavior of genes in the region, thus affecting the phenotype of the organism.

15. R. Durrett and D. Schmidt, "Waiting for regulatory sequences to appear," *Annals of Applied Probability* 17 (2007): 1–32. The relevant information appears on p. 19, where the time to fixation is factored in.

16. R. Durrett and D. Schmidt, "Waiting for two mutations: With applications to regulatory sequence evolution and the limits of Darwinian evolution," *Genetics 180 (2008): 1501–1509.*

17. A. K. Gauger *et al.*, "Reductive evolution can prevent populations from taking simple adaptive paths to high fitness," *BIO-Complexity* 1, no. 2 (2010): 1–9, doi:10.5048/BIO-C.

18. For a review pointing out unsolved conundrums concerning our uniqueness, see a recent review by A. Varki *et al.*, "Explaining human uniqueness: genome interactions with environment, behavior and culture," *Nature Reviews Genetics* 9 (2008): 749–763.

2

DARWIN'S LITTLE ENGINE

THAT COULDN'T

Douglas Axe

When it comes to producing major innovations in the history
of life like human beings, Darwin's engine of natural selection
acting on random variations has proved to be the little engine that
couldn't—certainly not in the time allowed by most scientists, and
probably not even in trillions of years.

BIOLOGIST RICHARD DAWKINS, A VOCAL ATHEIST, ONCE DESCRIBED biology as "the study of complicated things that give the appearance of having been designed for a purpose."[1] According to him, that appearance is entirely deceptive. Life needed no personal inventor because there is an impersonal one powerful enough to do the job, namely "[n]atural selection, the blind, unconscious, automatic process which Darwin discovered, and which we now know is the explanation for the existence and apparently purposeful form of all life,"[2] including us.

The evidence has convinced me otherwise. This engine of invention that Darwin imagined and Dawkins has spent much of his life promoting doesn't actually work very well when you put it to the test. I know this because I've been doing just that for a number of years, along with several of my colleagues. The results of our work have been described in technical detail. In fact, recognizing that the level of detail in these descriptions is far beyond what non-scientists are looking for, I'm going to focus here on the bigger picture that most interests us, and which has the added advantage of being amenable to communication in ordinary English.

The question of how we humans came to be—living, breathing things capable of pondering our own existence—is deeply connected to how we should think of ourselves. This places it among the most important subjects of human inquiry throughout the ages. Everyone perceives this, but when it comes to evaluating the science that gets drawn on to make arguments on this important topic, most people find themselves in the difficult position of having to judge a debate without speaking the language of the debaters. To add to the difficulty, the debaters themselves can be so emphatic and dogmatic that it seems as though something other than scientific data must be animating the exchange.

The good news is that the situation is not as hopeless as it might look. If careful observation and reasoning have anything decisive to say about our origin, then *science* provides a way forward. And by that I mean not any particular scientific authority or organization or committee or publication, but rather science itself. Science has always progressed by the conflict of ideas, and whatever benefit some of those ideas have received from things other than the twin pillars of observation and reasoning, those pillars alone will remain standing in the end. Every conclusion they don't support will fall... eventually.

With that in mind, my purpose here is to present a key part of the scientific case against Darwinism in terms that everyone can follow, and to tie that case to the great question of our own origin as humans. The best arguments are simple, so the very exercise of distilling an argument to its essence is, in my opinion, the best way for someone who labors over the technical details to step back and see whether anything good has come of it. I believe it has—that careful science now stands decisively against Darwinism. But whether you're inclined to agree or disagree, my aim is to equip you to decide for yourself.

DARWIN'S LITTLE ENGINE

MY COLLEAGUE Ann Gauger and I have recently challenged Darwin's engine to invent something so much simpler than humanity that the comparison may seem rather odd, and yet there is an important connection between what we examined and human origins. The technical details of our study are available for those who may want to examine them,[3] but all you need to know to follow what I'll say here is that each gene inside a cell carries the instructions for building a particular protein, and each protein is a tiny machine-like device that carries out one of the many tasks that must be accomplished for the cell to function properly.

Figure 2-1. A modest test that Darwin's engine failed. The object on the left is a depiction of the protein that we started with, and the object on the right is a protein that performs the desired new function. Keep in mind that we weren't asking whether the thing on the left can evolve the precise appearance of the thing on the right. We were simply asking whether it can evolve the function of the thing on the right. That should be possible with only a partial shift in the appearance, which ought to be relatively easy in view of the close resemblance.
Illustration: Douglas Axe.

In those simple terms, all we did was ask whether Darwin's engine can alter a single gene in bacterial cells so that its instructions specify a modified version of the original protein that performs a new task. We wanted this to work, so we bent over backwards to choose a pair of tasks that ought to make this conversion relatively easy. Since no one can predict how hard it might be to produce a protein function that's never been seen before, we did a thorough study of known proteins and chose a pair that are very similar but specialize in different tasks (the tasks themselves being similar in *kind*, but different in detail). In terms of more familiar objects, you can think of our test as being like taking a putter from the golf bag and asking something—some process—to reshape it to work as a pitching wedge. This is a real change of function, but not the fantastical kind that would be needed to get the wedge from a completely different object, like a corkscrew or a halogen lamp. Well, if the process in this example involves a talented metal worker then success is virtually guaranteed. But can something as simple and clueless as Darwin's engine really do anything comparable?

Apparently not, according to the results of our experiment. Darwin's engine proved to be the little engine that couldn't... certainly not in the few billion years in which it is supposed to have done everything, and probably not even in a few *trillion* years.

So, what does this have to do with our own origin? The answer is that it places an important limitation on what we can infer from similarity. Specifically, we now know that we *can't* infer that Darwin's engine can produce thing B from thing A simply because A and B are quite similar. We know this because we have now shown for a particular thing A and a similar thing B that his engine can't accomplish the transformation (not directly, anyway—more on that below). We also know in broad terms which aspect of our challenge caused the difficulty. It was that we required Darwin's engine to produce a new function. If we had been content for it to do something less, like modify the starting gene while preserving the function of its specified protein, then it would have passed. But that's like saying the loser of a contest would be thought a winner if we disregarded the contest. The failure of Darwin's engine in this case is its downfall precisely because we asked it to prove its most highly touted credentials—its credentials as an inventor.

It will be helpful to summarize our result in the form of a principle as follows:

Darwinian transitions from A to B that accomplish invention cannot be presumed plausible simply because A and B are substantially similar.

If principles can seem presumptuous when first introduced, the modesty of this one surely qualifies it as an exception. It doesn't say that *all* Darwinian transitions are implausible, like the one we studied. It simply says that their plausibility can't be counted on just because they end with something similar to what they started with.

Simple though it is, this principle turns out to have enormous implications for Darwinism. To fully grasp them, you have understand how central the concept of similarity has become to evolutionary reasoning. Since evolutionary biologists assume that Darwin's engine is capable of inventing everything that has been invented in the living world, their interest lies in the historical particulars of the engine's activity. They want to place life's key historical events on Darwin's *tree of life*, the great family tree he conceived as showing how all species descended from the first life. The general mechanism of invention itself no longer merits attention, this supposedly having been fully explained way back in the first half of the twentieth century when Darwin's theory was updated with the then current understanding of genetics. All that remains for today's evolutionary biologists is the business of inferring the details of the great family tree, and for this they need only continue the pursuit of methods for detecting the increasingly faint similarities left behind by increasingly distant familial relationships.

The logic of inference from similarity is very simple: *the greater the degree of similarity between two species, the closer their evolutionary relationship*. But notice what gets swept away by that simplicity. Having assumed that Darwin's engine can invent everything that got invented, biologists don't worry about whether the branches they propose in their attempts to reconstruct parts of the great tree are really plausible or not. The focus is entirely on whether similarities have been detected and grouped in a way that would convince other biologists, the thinking being that if those things are properly documented, then the evolutionary relationships inferred from them must be correct.

That turns out to be precarious reasoning. Considering that Darwin's engine operates through the ordinary process of procreation, if it really was the great inventor then all species are related in that ordinary procreative sense. But if we have reason to think it wasn't the great inventor, then the sense in which one species is related to the next must remain an open question until we settle the matter of the fundamental nature of the inventive process. As things now stand, our finding that a particular evolutionary transition between two very similar things is beyond the reach of Darwin's engine severely undermines the logic of similarity that has underwritten the entire Darwinian tree project. And now that *that* has been called into question, everything based on it needs to be reexamined.

A mental picture may help to clarify what went wrong. Darwinian evolution is often thought of in terms of journeys over a vast rugged landscape. Each point on this strange terrain represents a possible genome sequence, those possibilities being so staggeringly numerous that real organisms have only actualized a minute fraction of them. The ground elevation at each point corresponds to the fitness of individuals carrying that genome, with the horizontal distance between any two points indicating the degree to which the corresponding genomes differ. In terms of this picture, all of the millions of species alive today are represented by their own points, high up on peaks scattered somewhere across this conceptual landscape (the fact that they are *alive* demonstrates the quality of their genomes).

Now, wherever a species happens to be, Darwin's engine tends to move it toward the highest ground it can reach (**Figure 2-2**). According to the Darwinian story, that simple tendency to migrate upward has, over billions of years, transported the first primitive genome from its starting point to higher points along millions of diverging paths. The result is the spectacular variety of life forms we see today with a correspondingly wide dispersal of genomes across the vast conceptual landscape.

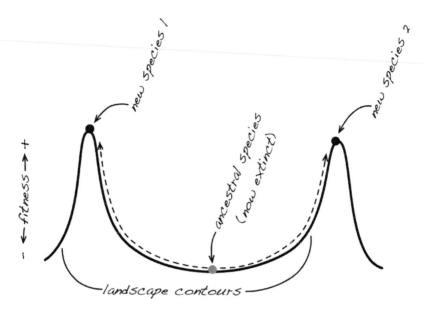

Figure 2-2. Darwin's explanation of the diversity of life forms. This is a cross-section through two peaks representing substantially different forms of life. The whole landscape stretches out in all directions, with millions of peaks representing all the different species. This simple hill-climbing mechanism, repeated million of times, is Darwin's explanation for the full variety of life.
Illustration: Douglas Axe.

But there's something suspicious about this story, as a number of careful observers pointed out long before Dr. Gauger and I did our experiment.[4] It has to do with the wide disparity of distance scales. The scale of the landscape, which is characterized by the extent to which dissimilar genomes differ, is very large by any reasonable calculation. On the other hand, Darwin's engine moves in steps that can only reach points a tiny distance away from the prior point. In one step it can move a genome to the highest point within this reach, but further progress would require a still higher point to fall within reach once that move is made. That might happen every now and then, but it would have to happen in an amazingly consistent and helpful way to explain how the enormous distances were traversed from the point marking the first primitive organism to the millions of points marking the great variety of modern life forms.

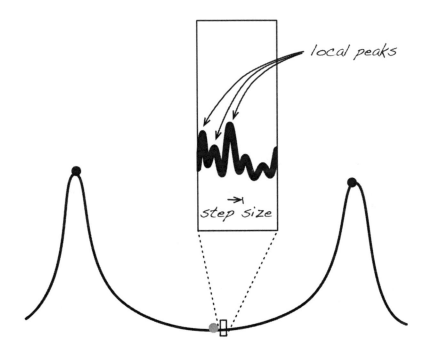

Figure 2-3. The problem of climbing in tiny steps. If the engine moves to the highest point that can be reached in each step and the landscape is rugged, then the endpoint will be a local peak.
Illustration: Douglas Axe.

Let's put this in more familiar terms. The summit of Mount Whitney, the highest point in the contiguous United States, is just 136 kilometers from the lowest point in North America, known as Badwater Basin. Now, suppose there were an automated vehicle capable of remotely scanning the surrounding terrain within some fixed distance and then moving to the highest point identified by the scan. If the scan radius is greater than 136 kilometers, this vehicle could get from Badwater to Whitney in one scan-and-move operation. But what if the scan radius is one *millionth* that size? Now the circle that the vehicle 'sees' from its current position is about a shoe-length across, with each move being up to half that distance. Considering how uneven the ground is, we wouldn't expect this nearsighted vehicle to complete more than a few scan-and-move operations before becoming stuck on a rock, maybe half a pace from where it started. Summiting Whitney would be completely out of the question. So the idea that *any* ability to seek higher ground, no matter how restricted, makes the highest summit accessible turns out to be highly simplistic.

The very same critique applies to Darwinism. Consider that for Darwin's engine to invent humans from apes, it would have had to work within the severe limitation of a single-mutation scan radius.[5] That is, it would have had to invent humans one simple mutation at a time, with each of these mutations making its possessors significantly more fit than their peers. Contrast this single-mutation reach with the *millions* of differences that distinguish the chimp and human genomes and we're back to the impossible trek from Badwater to Whitney. *Maybe* the genomic landscape is so much simpler and smoother than the Death Valley terrain as to enable Darwin's engine to cruise upward to exotic destinations on gentle inclines, but why would anyone assume this to be so? Only if experiment after experiment were to prove that remarkable kind of terrain to be the rule should anyone begin to think that something so fantastic might be true.

Alas, the experiment that we performed is one of many that have examined precisely this point, and the clear consensus is that the landscape is anything but smooth and gentle. We focused specifically on invention because this is where smoothness is crucial to the success of Darwin's evolutionary mechanism. In terms of the landscape picture, we placed Darwin's engine at a natural location (a genome with the gene for protein *A*) known to have a natural high peak that is very close to it (the same genome but with a gene for protein *B* instead). If that close peak had gently sloped faces, then the engine would have climbed it (as in **Figure 2-2**, but on a much smaller scale). It didn't. And while there are many examples in the scientific literature where the engine does climb for several steps, we have not found any where a new function was generated in the process. In fact, Darwin's engine often moves *away* from invention in its shortsighted pursuit of immediate fitness gains.[6]

Experiments will continue to add to this picture, of course. Darwin's engine can't drive the short distance from *A* to *B* in our test case, but perhaps an even smaller test will be found, and perhaps the engine will pass that test. For that matter, maybe a new protein will eventually be found that sits between our *A* and *B*, enabling the engine to traverse paths that connect *A* and *B* through that middle point. The important thing to realize, though, is that this wouldn't remove the problem of the disparity of scales. It is now clear that Darwin's engine can't climb a peak corresponding to a new invention unless that peak happens to be remarkably close to its current location—closer than the peak-to-peak distance between any pair of proteins that we know of with distinct functions. Even if such an extraordinary case were to be found, it would be just that—an extraordinary case. Traversing long distances would still depend on a very long and well coordinated succession of extraordinary cases, which amounts to nothing short of a miracle.

In fact, in my effort to simplify I've downplayed how extraordinary this would be. Darwin's engine actually received much more sympathetic treatment in our experiment than it would in nature. Most significantly, we made highly unrealistic arrangements in order for the intended conversion from *A* to *B* to be of any use to the bacterial cells we were working with. The truth is that several other conversions would have to occur and the whole set would have to come together by accident in one cell before anything of biological significance would happen. In terms of the landscape picture, that means the nearby peak we challenged Darwin's engine to climb would actually be much more distant in any realistic scenario.

THE VIEW FROM SAPIENS SUMMIT

WHEN IT comes to human origins, all I would add is that the inadequacy of Darwin's engine must surely become even more profound as the inventions attributed to it become more profound. Of the millions upon millions of amazing examples of invention to be seen among and within living things, none compares to the invention humanity—the invention of *inventors*. If the show on earth was spectacular when the likes of fireflies and geckos and orcas made their successive entrances, it became something incomparably deeper when humans took their place. Crickets brought more crickets, and chimps more chimps. All very good. Humans, on the other hand, brought the products of their own contemplation: music and drama and literature and painting and sculpture and philosophy and theology and mathematics and science and technology and athletics and culture and movements and politics and war. The best of good mixed with the worst of bad, all of it categorically unlike what came before—the chirping of crickets and the screeching of chimps.

So, if this humanity thing is on a level of its own, how reasonable do you suppose it is to chalk it up to Darwin's little engine? It's one thing to say that chimps and humans are similar enough that their likeness calls for careful explanation (few would argue with that), but as we've now seen it's quite another to say that they are similar enough for Darwin's engine to have traversed the gap between them. To insist on *that* is to ignore the evidence. A comparison of the complete human and chimp genomes has identified twenty distinct gene families, each with multiple genes, that are present in humans but absent from chimps and other mammals.[7] That's a huge gap when you compare it to the single in-family gene transition that we examined.

The truth is that humans have a tendency to accept what they've been told over and over, and scientists (being human) are no exception to this. Stories have their place in science, in the framing of ideas, but they aren't what makes good science so persuasive. So, scientists who insist that Darwin got our story, the human story, right would do well to ponder the evidence that would be needed to make that claim persuasive.

Have they thought seriously about what an ape-to-human transition would entail? Have they figured out how to wire a brain for speech, or for the intelligence needed to make use of speech? Do they know how to configure the lips, the tongue, and the vocal tract in order for speech to be physically possible? Have they discovered how to coordinate these inventions with all the changes needed for females to give birth to big-brained offspring?

And if they've mastered all these points while wearing their bioengineers' hats, have they switched to their geneticists' hats and identified a series of single mutations that would orchestrate this whole inventive process? They may think they know some of the answers to these problems, and that's a start, but have they gone into the primate lab and done the work that should convince those of us who wonder whether they have it right? Have they been hard at work for decades, quietly validating their ideas by producing talking chimps?

If so, have they done the experiments to measure the fitness effect of each single mutation along the line of chimps that eventually produced the ones that talk? Did they verify that each increases the fitness enough to become established in a natural population? And assuming they have checked all the boxes to this point, did they do the math to verify that the whole transition can happen naturally in an ape population within a few hundred thousand generations?

Hard questions are humbling, and humility may be the best way for scientists to earn the trust of their benefactors (the public) on this subject. In truth, almost nothing on the above checklist is technically feasible at present, so we don't need to lose any sleep over the ethical issues. My point is simply that virtually *everything* that would need to be done to establish the sheer physical possibility of turning apes into humans remains undone. And even in a strange sci-fi thought experiment where it *has* been done, the knowledge so gained would only further confirm how naive it is to think that Darwin's little engine could have done it.

A closing thought. As someone who loves science, I have to say that I can think of no conclusion in the whole history of the discipline that is so firm and so profound and so original that it should cause every human being to stop and rethink what it means to be human. Most simply aren't that profound. I happen to think that Darwin's *was* that profound, but thankfully, also profoundly wrong.

ENDNOTES

1. Richard Dawkins, *The Blind Watchmaker* (New York: Penguin, 1986), 1.
2. Ibid., 5.
3. A. K. Gauger and D. D. Axe, "The evolutionary accessibility of new enzyme functions: a case study from the biotin pathway," *BIO-Complexity* 2, no. 1 (2011): 1–17, accessed March 6, 2012, doi:10.5048/BIO-C.2011.1.
4. Paul S. Moorhead and Martin M. Kaplan, editors, *Mathematical Challenges to the Neo-Darwinian Interpretation of Evolution* (Philadelphia: Wistar Institute Press, 1967).
5. R. Durrett and D. Schmidt, "Waiting for two mutations: with applications to regulatory sequence evolution and the limits of Darwinian evolution," *Genetics* 180 (2008): 1501–1509, accessed March 6, 2012, doi:10.1534/genetics.107.082610.
6. A. K. Gauger, S. Ebnet, P. F. Fahey, and R. Seelke, "Reductive evolution can prevent populations from taking simple adaptive paths to high fitness," *BIO-Complexity* 1, no. 2 (2010): 1–9, accessed March 6, 2012, doi:10.5048/BIO-C.2010.2.
7. J. P. Demuth, T. De Bie, J. E. Stajich, N. Cristianini, and M. W. Hahn, "The evolution of mammalian gene families," *PLoS One* 1 (2006): e85, accessed March 6, 2012, doi:10.1371/journal.pone.0000085.

3

HUMAN ORIGINS AND THE

FOSSIL RECORD

Casey Luskin

> Hominin fossils generally fall into one of two groups: ape-like
> species and human-like species, with a large, unbridged gap
> between them. Despite the hype promoted by many evolutionary
> paleoanthropologists, the fragmented hominin fossil record does
> not document the evolution of humans from ape-like precursors.

EVOLUTIONARY SCIENTISTS COMMONLY TELL THE PUBLIC THAT THE fossil evidence for the Darwinian evolution of humans from ape-like creatures is incontrovertible. For example, anthropology professor Ronald Wetherington testified before the Texas State Board of Education in 2009 that human evolution has "arguably the most complete sequence of fossil succession of any mammal in the world. No gaps. No lack of transitional fossils... So when people talk about the lack of transitional fossils or gaps in the fossil record, it absolutely is not true. And it is not true specifically for our own species."[1] According to Wetherington, the field of human origins provides "a nice clean example of what Darwin thought was a gradualistic evolutionary change."[2]

Digging into the technical literature, however, reveals a story starkly different from the one presented by Wetherington and other evolutionists engaging in public debates. As this chapter will show, the fossil evidence for human evolution remains fragmentary, hard to decipher, and hotly debated.

Indeed, far from supplying "a nice clean example" of "gradualistic evolutionary change," the record reveals a dramatic discontinuity between ape-like and human-like fossils. Human-like fossils appear abruptly in the record, without clear evolutionary precursors, making the case for human evolution based on fossils highly speculative.

THE CHALLENGES OF PALEOANTHROPOLOGY

HUMANS, CHIMPS, and all of the organisms leading back to their supposed most recent common ancestor are classified by evolutionary scientists as "hominins." The discipline of paleoanthropology is devoted to the study of the fossil remains of ancient hominins. Paleoanthropologists face a number of daunting challenges in their quest to reconstruct a story of hominim evolution.

First, **hominin fossils tend to be few and far between.** It's not uncommon for long periods of time to exist for which there are few fossils documenting the evolution that was supposedly taking place. As paleoanthropologists Donald Johanson (the discoverer of Lucy) and Blake Edgar observed in 1996, "[a]bout half the time span in the last three million years remains undocumented by any human fossils" and "[f]rom the earliest period of hominid evolution, more than 4 million years ago, only a handful of largely undiagnostic fossils have been found."[3] So "fragmentary" and "disconnected" is the data that in the judgment of Harvard zoologist Richard Lewontin, "no fossil hominid species can be established as our direct ancestor."[4]

The second challenge faced by paleoanthropologists is the **fossil specimens themselves.** Typical hominin fossils consist literally of mere bone fragments, making it difficult to make definitive conclusions about the morphology, behavior, and relationships of many specimens. As the late paleontologist Stephen Jay Gould noted, "[m]ost hominid fossils, even though they serve as a basis for endless speculation and elaborate storytelling, are fragments of jaws and scraps of skulls."[5]

A third challenge is **accurately reconstructing the behavior, intelligence, or internal morphology of extinct organisms.** Using an example from living primates, primatologist Frans de Waal observes that the skeleton of the common chimpanzee is nearly identical to its sister species, the bonobo, but they have great differences in behavior. "On the sole basis of a few bones and skulls," writes de Waal, "no one would have dared to propose the dramatic behavioral differences recognized today between the bonobo and the chimpanzee."[6] He argues this should serve as "a warning for paleontologists who are reconstructing social life from fossilized remnants of long-extinct species."[7] De Waal's example pertains to a case where the investigators have complete skeletons, but the late University of Chicago anatomist C. E. Oxnard explained how these problems are intensified when bones are missing: "A series of associated foot bones from Olduvai [a locality bearing australopithecine fossils] has been reconstructed into a form closely resembling the human foot today although a similarly incomplete foot of a chimpanzee may also be reconstructed in such a manner."[8]

Flesh reconstructions of extinct hominins are likewise often highly subjective. They may attempt to diminish the intellectual abilities of humans and overstate of those of animals. For example, one popular high school textbook[9] caricatures *Neanderthals* as intellectually primitive even though they exhibited signs of art, language, and culture,[10] and casts *Homo erectus* as a bungling, stooped form even though its postcranial skeleton is extremely similar to that of modern humans.[11] Conversely, the same textbook portrays an ape-like australopithecine with gleams of human-like intelligence and emotion in its eyes—a tactic common in illustrated books on human origins.[12] University of North Carolina, Charlotte anthropologist Jonathan Marks warns against this when lamenting the "fallacies" of "humanizing apes and ape-ifying humans."[13] The words of the famed physical anthropologist Earnest A. Hooton from Harvard University still ring true: "alleged restorations of ancient types of man have very little, if any, scientific value and are likely only to mislead the public."[14]

Given these challenges, one might expect caution, humility, and restraint from evolutionary scientists when discussing hypotheses about human origins. And sometimes this is indeed found. But as multiple commentators have recognized, we often find precisely the opposite.[15]

Calm and collected scientific objectivity in the field of evolutionary paleoanthropology can be as rare as the fossils themselves. The fragmented nature of the data, combined with the desire of paleoanthropologists to make confident assertions about human evolution, leads to sharp disagreements within the field, as pointed out by Constance Holden in her article in *Science* titled "The Politics of Paleoanthropology."

Holden acknowledges that "[t]he primary scientific evidence" relied on by paleoanthropologists "to construct man's evolutionary history" is "a pitifully small array of bones... One anthropologist has compared the task to that of reconstructing the plot of *War and Peace* with 13 randomly selected pages."[16] According to Holden, it is precisely because researchers must draw their conclusions from this "extremely paltry evidence" that "it is often difficult to separate the personal from the scientific disputes raging in the field."[17]

Make no mistake: The disputes in paleoanthropology are often deeply personal. As Donald Johanson and Blake Edgar admit, ambition and lifelong quests for recognition, funding, and fame, can make it difficult for paleoanthropologists to admit when they are wrong: "The appearance of discordant evidence is sometimes met with a sturdy reiteration of our original views... it takes time for us to give up pet theories and assimilate the new information. In the meantime, scientific credibility and funding for more fieldwork hang in the balance."[18]

Indeed, the quest for recognition can inspire outright contempt toward other researchers. After interviewing paleoanthropologists for a documentary in 2002, PBS NOVA producer Mark Davis reported that "[e]ach Neanderthal expert thought the last one I talked to was an idiot, if not an actual Neanderthal."[19]

It's no wonder that paleoanthropology is a field rife with dissent and with few universally accepted theories among its practitioners. Even the most established and confidently asserted theories of human origins may be based upon limited and incomplete evidence. In 2001, *Nature* editor Henry Gee conceded, "[f]ossil evidence of human evolutionary history is fragmentary and open to various interpretations."[20]

The Standard Story of Human Evolutionary Origins

Despite the widespread disagreements and controversies just described, there is a standard story of human origins which is retold in countless textbooks, newsmedia articles, and coffee table books. A representation of the most commonly believed hominin phylogeny is portrayed below in **Figure 3-1**:

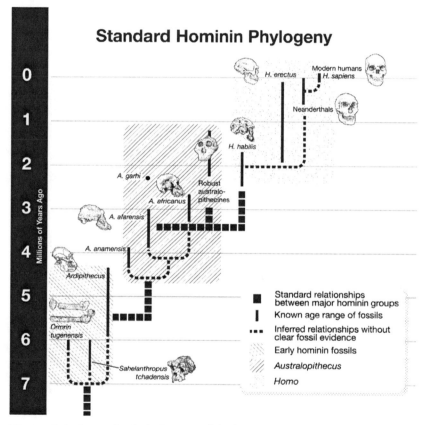

Figure 3-1: A standard phylogeny of the branch of the hominin tree that includes humans.[21]
Illustration: Jonathan Jones.

Starting with the early hominins at the bottom left, and moving upwards through the australopithecines, and then into members of the genus *Homo*, this chapter will review the fossil evidence and assess whether it supports this alleged story of human evolution. As we shall see, the evidence—or lack thereof—often gets in the way of the evolutionary story.

EARLY HOMININ FOSSILS

ALTHOUGH GIVEN much hype in the media, the earliest hominin fossils are often so fragmentary that they remain the subject of considerable controversy in the scientific community. This section will examine some of the main early hominin fossils and the debates surrounding them.

Sahelanthropus tchadensis: "Toumai Skull"

Despite the fact that *Sahelanthropus tchadensis* (also called the "Toumai skull") is known only from one skull and some jaw fragments, it has been called the oldest known hominin that lies directly on the human line.

But not everyone agrees. When the fossil was first reported, Brigitte Senut, a leading researcher at the Natural History Museum in Paris, said "I tend towards thinking this is the skull of a female gorilla."[22] Writing in *Nature* with Milford H. Wolpoff, Martin Pickford, and John Hawks, Senut later noted there are "many… features that link the specimen with chimpanzees, gorillas or both, to the exclusion of hominids," and argued "*Sahelanthropus* does not appear to have been an obligate biped."[23] In their view, "*Sahelanthropus* was an ape."[24]

This debate has continued, but leading paleoanthropologists have cautioned in the *Proceedings of the National Academy of Sciences (USA)* that teeth and skull fragments alone are insufficient to properly classify or understand species as a hominin: "[O]ur results show that the type of craniodental characters that have hitherto been used in hominin phylogenetics are probably not reliable for reconstructing the phylogenetic relationships of higher primate species and genera, including those among the hominins."[25]

At one point during the Texas evolution hearings, Ronald Wetherington testified that "every fossil we find reinforces the sequence that we had previously supposed to exist rather than suggesting something different."[26] But this fossil, first reported in 2002, provides a striking counterexample to that assertion. Commenting on the Toumai skull in the journal *Nature*, Bernard Wood of George Washington University opened by observing, "A single fossil can fundamentally change the way we reconstruct the tree of life."[27] He went on to state:

> If we accept these as sufficient evidence to classify *S. tchadensis* as a hominid at the base, or stem, of the modern human clade, then it plays havoc with the tidy model of human origins. Quite simply, a hominid of this age should only just be beginning to show signs of being a hominid. It certainly should not have the face of a hominid less than one-third of its geological age. Also, if it is accepted as a stem hominid, under the tidy model the principle of parsimony dictates that all creatures with more primitive faces (and that is a very long list) would, perforce, have to be excluded from the ancestry of modern humans.[28]

In other words, if the Toumai skull is accepted as a stem ancestor of humans, then many later supposed human ancestors—including the acclaimed australopithecines—ought not be considered human ancestors. Wood concludes that fossils like *Sahelanthropus* show "compelling evidence that our own origins are as complex and as difficult to trace as those of any other group of organisms."[29]

Orrorin tugenensis: "**Orrorin**"

Orrorin, which means "original man" in a local Kenyan language, was a chimpanzee-sized primate which is known only from "an assortment of bone fragments,"[30] including pieces of the arm, thigh, and lower jaw, as well as some teeth (**Figure 3-2**). When initially discovered, the *New York Times* ran a story titled "Fossils May Be Earliest Human Link,"[31] and reported it "may be the earliest known ancestor of the human family."[32] Despite the meagerness of the find, enough enthusiasm was stirred that an article in *Nature* soon after the fossil's unveiling cautioned that "excitement needs to be tempered with caution in assessing the claim of a six-million-year-old direct ancestor of modern humans."[33]

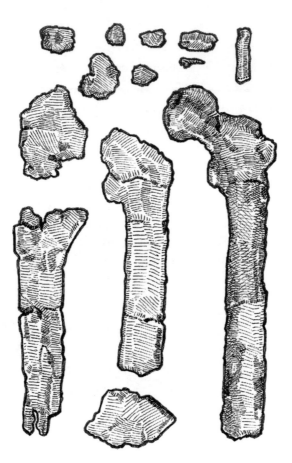

Figure 3-2: Fragments of *Orrorin tugenensis*.
Illustration: Jonathan Jones.

Some paleoanthropologists claimed that *Orrorin's* femur indicates
a bipedal mode of locomotion which was "appropriate for a population
standing at the dawn of the human lineage."[34] But as a later Yale University Press commentary admitted, "All in all, there is currently precious
little evidence bearing on how *Orrorin* moved."[35]

Evolutionary paleoanthropologists often assume that bipedality is
a litmus test for membership along the human line. So if *Orrorin* did
prove to be an upright-walking ape-like creature from six million years

ago (mya), would that qualify it as a human ancestor? Not at all. In fact, the fossil record contains bipedal apes which evolutionists recognize were far removed from the human line. In 1999, UC San Diego biologist Christopher Wills observed that "[u]pright posture may not be unique to our own lineage" since "[a]n ape that lived ten million years ago on Sardinia, *Oreopithecus bambolii*, seems to have acquired similar capabilities, perhaps independently."[36] A more recent article in ScienceDaily elaborated:

> *Oreopithecus bambolii*, a fossil ape from Italy shares many similarities with early human ancestors, including features of the skeleton that suggest that it may have been well adapted for walking on two legs. However, the authors observe, enough is known of its anatomy to show that it is a fossil ape that is only distantly related to humans, and that it acquired many "human-like" features in parallel.[37]

A 2011 paper in *Nature* by Bernard Wood and Terry Harrison explains the implications of bipedal apes that had nothing to do with human origins:

> The object lesson that *Oreopithecus* provides is critical to the debate about interpreting the relationships of the earliest purported hominins. It demonstrates how features considered to be hominin specializations can be shown to have been acquired independently in a non-hominin lineage in association with inferred behaviours that are functionally related to, but not necessarily narrowly restricted to, terrestrial bipedalism.[38]

Much as the Toumai skull threatened to displace australopithecines from our ancestral line, Pickford and his co-authors argued that if their hypothesis about *Orrorin* is correct, then australopithecines are again no longer ancestral to humans, but were merely "a side branch of hominid evolution that went extinct."[39] This hypothesis was not well-received by the paleoanthropological community, because they need the australopithecines to serve as an evolutionary precursor leading to our genus *Homo*. Another paper in *Nature* exemplified how dissenting views are treated in paleoanthropology, charging that Pickford's "simple phylogeny contrasts starkly with mainstream ideas about human evolution, and glosses over many areas of controversy and uncertainty."[40]

While *Orrorin* offers evolutionary paleoanthropologists the tantalizing possibility of a bipedal creature that lived around the time of the supposed split between humans and chimpanzees, simply too little of it is known at present to make confident claims about its locomotion, or its proper place in the supposed evolutionary tree.

Ardipithecus ramidus: "Ardi"

In 2009, the journal *Science* announced the publication of long-awaited reports about a 4.4 mya fossil named *Ardipithecus ramidus*. Expectations were high as its discoverer, UC Berkeley paleoanthropologist Tim White, had previously promised the fossil was a "phenomenal individual" that would be the "Rosetta stone for understanding bipedalism."[41] When the papers were finally released, the science media took it as an opportunity to evangelize the public for Darwin via the fossil they affectionately dubbed "Ardi."

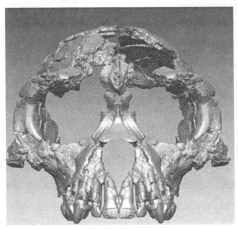

Figure 3-3: Anterior view of fragmented and reconstructed *Ardipithecus ramidus* skull.

Illustration: Used with permission from American Association for the Advancement of Science, Figure 2(D), Gen Suwa, Berhane Asfaw Reiko T. Kono, Daisuke Kubo4, C. Owen Lovejoy, and Tim D. White, "The Ardipithecus ramidus Skull and Its Implications for Hominid Origins," 326 (October 2, 2009): 68e1–68e7. Copyright 2009.

The Discovery Channel ran the headline "'Ardi,' Oldest Human Ancestor, Unveiled," and quoted White stating Ardi is "as close as we have ever come to finding the last common ancestor of chimpanzees and humans."[42] The Associated Press's headline read "World's oldest human-linked skeleton found," and stated "the new find provides evidence that chimps and humans evolved from some long-ago common ancestor."[43] *Science* magazine named Ardi the "breakthough of the year" for 2009,[44] and officially introduced her with an article titled "A New Kind of Ancestor: *Ardipithecus* Unveiled."[45] (A reconstruction of Ardi's skull can be seen in **Figure 3-3**.)

Calling this fossil "new" may have been a poor word choice on the part of *Science*, since Ardi was discovered in the early 1990s. Why did it take over 15 years for reports to be published? A 2002 article in *Science* explains that initially the bones were so "soft," "crushed," "squished," and "chalky," that White reported, "when I clean an edge it erodes, so I have to mold every one of the broken pieces to reconstruct it."[46]

Later reports similarly acknowledged that "some portions of Ardi's skeleton were found crushed nearly to smithereens and needed extensive digital reconstruction," and that its pelvis initially "looked like an Irish stew."[47] The 2009 report in *Science* tells a striking story of the fossil's poor quality: "[T]he team's excitement was tempered by the skeleton's terrible condition. The bones literally crumbled when touched. White called it road kill. And parts of the skeleton had been trampled and scattered into more than 100 fragments; the skull was crushed to 4 centimeters in height."[48] In an article titled "Oldest Skeleton of Human Ancestor Found," the science editor at *National Geographic*, put it this way: "After Ardi died, her remains apparently were trampled down into mud by hippos and other passing herbivores. Millions of years later, erosion brought the badly crushed and distorted bones back to the surface. They were so fragile they would turn to dust at a touch."[49]

Claims about bipedal locomotion in hominids require careful and accurate measurements of the precise shape of various bones. How far should one trust claims about Ardi as a "Rosetta stone for understanding bipedalism" when the bones were initially "crushed to smithereens" and "would turn to dust at a touch"? Several skeptical paleoanthropologists felt those claims warranted little credence. As *Science* reported:

> [S]everal researchers aren't so sure about these inferences. Some are skeptical that the crushed pelvis really shows the anatomical details needed to demonstrate bipedality. The pelvis is "suggestive" of bipedality but not conclusive, says paleoanthropologist Carol Ward of the University of Missouri, Columbia. Also, *Ar. ramidus* "does not appear to have had its knee placed over the ankle, which means that when walking bipedally, it would have had to shift its weight to the side," she says. Paleoanthropologist William Jungers of Stony Brook University in New York state is also not sure that the skeleton was bipedal. "Believe me, it's a unique form of bipedalism," he says. "The postcranium alone would not unequivocally signal hominin status, in my opinion."[50]

A subsequent paper by primatologist Esteban Sarmiento in *Science* noted that "All of the *Ar. ramidus* bipedal characters cited also serve the mechanical requisites of quadrupedality, and in the case of *Ar. ramidus* foot-segment proportions, find their closest functional analog to those of gorillas, a terrestrial or semiterrestrial quadruped and not a facultative or habitual biped."[51]

Critics also questioned the claim that Ardi was necessarily ancestral to humans. When Ardi's reports were first published, Bernard Wood stated, "I think the head is consistent with it being a hominin... but the rest of the body is much more questionable."[52] Two years later, Wood cowrote a paper in *Nature* elaborating on those criticisms, observing that if "*Ardipithecus* is assumed to be a hominin," and ancestral to humans, then this implies the fossil had "remarkably high levels of homoplasy among extant great apes."[53] In other words, Ardi had many ape-like characteristics which, if we set aside the preferences of many evolutionary paleoanthropologists, might imply a much closer relationship to living apes than to humans. According to a *ScienceDaily* article reporting on Wood's *Nature* paper, the claim of Ardi "being a human ancestor is by no means the simplest, or most parsimonious explanation."[54] Stanford University anthropologist Richard Klein put it this way: "I frankly don't think Ardi was a hominid, or bipedal."[55]

Sarmiento observed that Ardi had characters which were different not just from humans, but also from apes. In a *Time Magazine* interview titled, "Ardi: The Human Ancestor Who Wasn't?," he elaborated:

> "[Tim White] showed no evidence that Ardi is on the human lineage," Sarmiento says. "Those characters that he posited as relating exclusively to humans also exist in apes and ape fossils that we consider not to be in the human lineage."
>
> The biggest mistake White made, according to the paper, was to use outdated characters and concepts to classify Ardi and to fail to identify anatomical clues that would rule her out as a human ancestor. As an example, Sarmiento says that on the base of Ardi's skull, the inside of the jaw joint surface is open as it is in orangutans and gibbons, and not fused to the rest of the skull as it is in humans and African apes—suggesting that Ardi diverged before this character developed in the common ancestor of humans and apes.[56]

Whatever Ardi may have been, everyone agrees that this fossil was initially badly crushed and needed extensive reconstruction. Its discoverers adamantly maintain the specimen was a bipedal human ancestor, or something very close to it. No doubt this debate will continue, but are we obligated to take for granted the bold talking points promoted by Ardi's discoverers in the media? Sarmiento doesn't think so. According to *Time Magazine*, he "regards the hype around Ardi to have been overblown."[57]

LATER HOMININS: THE AUSTRALOPITHECINES

IN APRIL 2006, *National Geographic* ran a story titled "Fossil Find Is Missing Link in Human Evolution, Scientists Say,"[58] which reported the discovery of what the Associated Press called "the most complete chain of human evolution so far."[59] The fossils, belonging to the species *Australopithecus anamensis* were said to link *Ardipithecus* to its supposed australopithecine descendants.

What exactly was found? According to the technical paper reporting the find, the bold claims were based upon a few fragmented canine teeth which were said to be "intermediate" in size and shape. The technical description used in the paper was intermediate "masticatory robusticity."[60] If a couple of four million-year-old teeth of "intermediate" size and shape make "the most complete chain of human evolution so far," then the evidence for human evolution must be indeed quite modest.

Besides learning to distrust media hype, there is another important lesson to be gained from this episode. Accompanying the praise of this "missing link" were what might be called retroactive confessions of ignorance. In this common phenomenon, evolutionists acknowledge a severe gap in their evolutionary claims only after they think they have found evidence to plug that gap. Thus, the technical paper that reported these teeth admitted that, "Until recently, the origins of *Australopithecus* were obscured by a sparse fossil record,"[61] further stating: "The origin of *Australopithecus*, the genus widely interpreted as ancestral to *Homo*, is a central problem in human evolutionary studies. *Australopithecus* species differ markedly from extant African apes and candidate ancestral hominids such as *Ardipithecus, Orrorin* and *Sahelanthropus*."[62] Following these comments, an article on MSNBC.com acknowledged that "Until now, what scientists had were snapshots of human evolution scattered around the world."[63]

Evolutionists who make retroactive confessions of ignorance risk the danger that the evidence which supposedly filled the gap may turn out to not be so compelling after all. This seems to be the case here, where a couple teeth of intermediate "masticatory robusticity" were apparently all that stood between an unsolved "central problem in human evolutionary studies," and "the most complete chain of human evolution so far."

Moreover, we're left with the uncontested admission that the australopithecines "differ markedly" from their supposed ancestors—*Ardipithecus, Orrorin,* or *Sahelanthropus.* Given the fragmentary and enigmatic nature of those earlier species, a more objective analysis might suspect that this period of supposed early hominin evolution remains what Tim White once called it: "a black hole in the fossil record."[64]

Australopithecines Are Like Apes

While *Sahelanthropus, Orrorin,* and *Ardipithecus* are controversial due to the fragmented nature of their remains, there are sufficient known specimens of the australopithecines to gain a better understanding of their morphology. Nonetheless, controversy remains over whether the australopithecines were upright-walking ancestors of the genus *Homo.*

Australopithecus, which literally means "southern ape," is a group of extinct hominins that lived in Africa from a little over 4 mya until about 1 mya. "Splitters" (those paleoanthropologists who tend to see many different species in the fossil record) and "lumpers" (those who see fewer) have created a variety of taxonomic schemes for the australopithecines. However, the four most commonly accepted species are *afarensis, africanus, robustus,* and *boisei. Robustus* and *boisei* are larger boned and more "robust" than the others and are sometimes classified under the genus *Paranthropus.*[65] According to conventional evolutionary thinking, they represent a later-living offshoot that went extinct without leaving any living descendants today. The smaller "gracile" forms, *africanus* and *afarensis* (the species which includes the famous fossil "Lucy"), lived earlier, and are classified within the genus *Australopithecus.* These two latter species are commonly said to be directly ancestral to humans.

By far, the most well-known australopithecine fossil is Lucy because she is one of the most complete fossils among known pre-*Homo* hominins. She is commonly claimed to have been a bipedal ape-like creature which serves as an ideal precursor to the human species.

In 2009, Lucy's skeleton came to the Pacific Science Center in my hometown of Seattle. Upon entering the room containing the thick glass case holding her bones, I was immediately struck by the incompleteness of her skeleton. Only 40% was found, and a significant percentage is mere rib fragments. (See **Figure 3-4**.) Very little useful material from

Lucy's skull was recovered, and yet she is one of the most significant specimens ever found.

Figure 3-4: The skeletal remains of "Lucy."
Illustration: Redrawn by Jonathan Jones based on Wikimedia Commons image of Lucy skeleton licensed under Creative Commons Attribution–Share Alike 3.0 Unported license.

There are some reasons for skepticism over whether "Lucy" represents a single individual, or even from a single species. In a video playing at the exhibit, Lucy's discoverer Donald Johanson admitted that when he found the fossil, the bones were scattered across a hillside, where he "looked up the slope and there were other bones sticking out." Johanson's written account explains further how the bones were not found together: "[S]ince the fossil wasn't found *in situ*, it could have come from anywhere above. There's no matrix on any of the bones we've found either. All you can do is make probability statements."[66]

This was therefore not a case where the bones were found connected forming a contiguous skeleton, but rather they were scattered across a hillside. Ann Gibbons notes that Johanson's "entire team fanned out over the gully to collect Lucy's bones."[67] At one point, Johanson explains that if there had been only one more rainstorm, Lucy's bones might have been washed away, never to be seen again. This does not inspire confidence in the integrity of the skeleton: If the next rainstorm could have washed Lucy away completely, what might have happened during prior storms to mix her up with who-knows-what? Could "Lucy" represent bones from multiple individuals or even multiple species?

The classical rejoinder notes that none of Lucy's bones appear duplicated, implying they come from a single individual. This is certainly possible, but given the fragmented and the incomplete and scattered nature of the skeleton, the rebuttal argument is far from conclusive. In particular, it's difficult to say with high confidence that key portions of the skeleton—such as the half-pelvis and half-femur—are from the same individual. The pelvis and femur are, after all, her most studied bones, and are said to indicate she walked upright. As the Pacific Science Center exhibit boldly stated, "Lucy's species walked bipedally, in much the same way as we do," at one point claiming her skeleton "approximate[s] a chimpanzee-like head perched atop a human-like body."

Lucy *did* have a small, chimp-like head in both size and shape—as University of Witwatersrand paleoanthropologist Lee Berger observes, "Lucy's face would have been prognathic, jutting out almost to the same degree as a modern chimpanzee."[68] But many have disagreed with claims that she looked like an ape-human hybrid. Bernard Wood refutes this misapprehension: "Australopithecines are often wrongly thought to have had a mosaic of modern human and modern ape features, or, worse, are regarded as a group of 'failed' humans. Australopithecines were neither of these."[69]

Moreover, many have challenged the claim that Lucy walked like we do, or was even significantly bipedal. Mark Collard and Leslie Aiello observe in *Nature* that much of the rest of her body was "quite ape-like," especially with respect to the "relatively long and curved fingers, relatively long arms, and funnel-shaped chest."[70] Their article also reports "good evidence" from Lucy's hand-bones that her species "'knuckle-walked', as chimps and gorillas do today."[71]

Needless to say, paleoanthropologists who wish Lucy to be a bipedal precursor to our genus *Homo* disfavor the "knuckle-walking" interpretation. Collard and Aiello fall into this category, calling this evidence "counterintuitive," and suggesting that "the locomotor repertoire of *A. afarensis* included forms of bipedalism, climbing and knuckle-walking." This proposal is tenuous, however, since these forms of locomotion tend to be mutually exclusive. Nonetheless, they dismiss Lucy's knuckle-walking specializations as "primitive retentions" from her ancestors.[72] Science writer Jeremy Cherfas explains why this argument is doubtful:

> Everything about her skeleton, from fingertips to toes, suggests that Lucy and her sisters retain several traits that would be very suitable for climbing in trees. Some of those same treeclimbing adaptations can still be detected, albeit much reduced, in much later hominids such as the 2-million-year old specimens of *Homo habilis* from the Olduvai gorge. It could be argued that Lucy's arboreal adaptations are just a hangover from her treedwelling past, but animals do not often retain traits that they do not use, and to find those same features in specimens 2 million years later makes it most unlikely that they are remnants.[73]

Apparently when the evidence points against Lucy being bipedal, it is simply discarded. But the main motivation for this dismissal is the evolutionary belief that modern humans need fully bipedal ape-like ancestors.

Other leading paleoanthropologists also acknowledge that Lucy's mode of locomotion was significantly different from that of humans. Richard Leakey and Roger Lewin argue that *A. afarensis* and other australopithecines "almost certainly were not adapted to a striding gait and running, as humans are."[74] Their quotation of anthropologist Peter Schmid's surprise at the non-human qualities of Lucy's skeleton is striking:

"We were sent a cast of the Lucy skeleton, and I was asked to assemble it for display," remembers Peter Schmid, a paleontologist at the Anthropological Institute in Zurich... "When I started to put [Lucy's] skeleton together, I expected it to look human," Schmid continues. "Everyone had talked about Lucy as being very modern, very human, so I was surprised by what I saw"... "What you see in *Australopithecus* is not what you'd want in an efficient bipedal running animal," says Peter. "The shoulders were high, and, combined with the funnel-shaped chest, would have made arm swinging very improbable in the human sense. It wouldn't have been able to lift its thorax for the kind of deep breathing that we do when we run. The abdomen was potbellied, and there was no waist, so that would have restricted the flexibility that's essential to human running."[75]

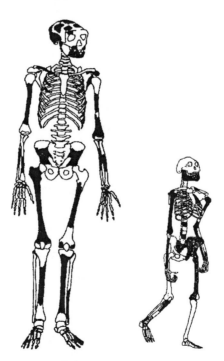

Figure 3-5: A comparison of *Australopithecus* (right) to early *Homo* (left). Black bones indicate those which have been discovered.[76]

Illustration: From Figure 1. John Hawks et. al., "Population Bottlenecks and Pleistocene Human Evolution," Molecular Biology and Evolution, copyright 2000, 17 (1): 2–22, by permission of the Society for Molecular Biology and Evolution.

Other studies confirm australopithecine differences with humans, and similarities with apes. Their inner ear canals—responsible for balance and related to locomotion—are different from those of *Homo* but similar to those of great apes.[77] Their ape-like developmental patterns[78] and ape-like ability for prehensile grasping by their toes[79] led one reviewer in *Nature* to say that whether australopithecines "were phylogenetically hominines or not, it seems to me that ecologically they may still be considered as apes."[80] In 1975 C. E. Oxnard published a paper in *Nature* using multivariable statistical analysis to compare key australopithecine skeletal characteristics to living hominids. He found that australopithecines have "a mosaic of features unique to themselves and features bearing some resemblances to those of the orangutan" and concluded: "If these estimates are true, then the possibility that any of the australopithecines is a direct part of human ancestry recedes."[81] Even the teeth of Lucy's species have been found to conflict with the hypothesis she was a human ancestor. A 2007 paper in *Proceedings of the National Academy of Sciences (USA)* reported "[g]orilla-like anatomy on *Australopithecus afarensis* mandibles," which was "unexpected," and "cast[s] doubt on the role of *Au. afarensis* as a modern human ancestor."[82]

As for Lucy's pelvis, many have claimed it supports a bipedal form of locomotion, but Johanson and his team reported it was "badly crushed" with "distortion" and "cracking" when first discovered.[83] These problems led one commentator to propose in the *Journal of Human Evolution* that the reason Lucy's pelvis is "so different from other australopithecines and so close to the human condition" was "error in the reconstruction... creating a very 'human-like' sacral plane."[84] Another paper in the same journal concluded that the lack of clear fossil data about Lucy prevents paleoanthropologists from making firm conclusions about her mode of locomotion: "Prevailing views of Lucy's posture are almost impossible to reconcile... To resolve such differences, more anatomical (fossil) evidence is needed. The available data at present are open to widely different interpretations."[85]

Paleoanthropologist Leslie Aiello, who served as head of the anthropology department at University College London, states that when it comes to locomotion, "[a]ustralopithecines are like apes, and the *Homo* group are like humans. Something major occurred when *Homo* evolved, and it wasn't just in the brain."[86]

The "something major" that occurred was the abrupt appearance of the human body plan—without direct evolutionary precursors in the fossil record.

A Big Bang Theory of Homo

IF HUMAN beings evolved from ape-like creatures, what were the transitional species between the ape-like hominins just discussed and the truly human-like members of the *Homo* genus found in the fossil record?

There aren't any good candidates.

Many paleoanthropologists have cited *Homo habilis*, dated at about 1.9 mya,[87] as a transitional species between the australopithecines and our genus *Homo*. But there are many questions about what exactly habiline specimens were. In the words of Ian Tattersall, an anthropologist at the American Museum of Natural History, the species is "a wastebasket taxon, little more than a convenient recipient for a motley assortment of hominin fossils."[88] As recent as 2009, Tattersall reaffirmed this view, writing with Jeffrey Schwartz that *habilis* represents "a rather heterogeneous assemblage, and it is probable that more than one hominid species is represented."[89]

Penn State University paleoanthropologist Alan Walker explains the severity of disagreements over this species: "[T]his is not a matter of some fragmentary fossils that are difficult to agree on. Whole crania are placed by different people in different species or even genera."[90] One reason for the disagreements is that the quality of the fossils is often poor. As Walker puts it, "[d]espite the number of words published on this species... there is not as much bony evidence as we would like."[91]

Ignoring these difficulties and assuming that *H. habilis* existed as the species many claim it was, there is a chronological consideration which precludes it from being ancestral to *Homo*. Habiline remains do not predate the earliest fossil evidence of true members of *Homo*, which appear about 2 mya. As a consequence, *H. habilis* could not have been a precursor to our genus.[92]

Morphological analyses further confirm that *habilis* makes an unlikely candidate as an "intermediate" or "link" between *Australopithecus* and *Homo*. An authoritative review paper titled "The Human Genus," published in *Science* in 1999 by leading paleoanthropologists Bernard Wood and Mark Collard found that *habilis* is different from *Homo* in terms of body size, body shape, mode of locomotion, jaws and teeth, developmental patterns, and brain size, and should be reclassified within *Australopithecus*.[93] A 2011 article in *Science* similarly noted that *habilis* "matured and moved less like a human and more like an australopithecine," had a dietary range "more like Lucy's than that of *H. erectus*."[94] Like the australopithecines, many features of *habilis* indicate they were more similar to modern apes than humans. According to Wood, habilines "grew their teeth rapidly, like an African ape, in contrast to the slow dental development of modern humans."[95]

An analysis in *Nature* of the ear canals of *habilis* similarly found that its skull is most similar to baboons and suggested the fossil "relied less on bipedal behaviour than the australopithecines."[96] The article concluded that "[p]hylogenetically, the unique labyrinth of [the *habilis* skull] represents an unlikely intermediate between the morphologies seen in the australopithecines and *H. erectus*."[97] Additionally, a study by Sigrid Hartwig-Scherer and Robert D. Martin in the *Journal of Human Evolution* found that the skeleton of *habilis* was *more* similar to living apes than were other australopithecines like Lucy.[98] They concluded: "It is difficult to accept an evolutionary sequence in which *Homo habilis*, with less human-like locomotor adaptations, is intermediate between *Australopithecus afaren[s]is* ... and fully bipedal *Homo erectus*."[99] Elsewhere, Hartwig-Scherer explained "expectations concerning postcranial similarities between *Homo habilis* and later member of the genus *Homo* could not be corroborated."[100]

To the contrary, she explains, *habilis* "displays much stronger similarities to African ape limb proportions," than even Lucy.[101] She called these results "unexpected in view of previous accounts of *Homo habilis* as a link between australopithecines and humans."[102]

Without *habilis* as an intermediate, it is difficult to find fossil hominins to serve as direct transitional forms between the australopithecines and *Homo*. Rather, the fossil record shows dramatic and abrupt changes which correspond to the appearance of *Homo*.

A 1998 article in *Science* noted that at about 2 mya, "cranial capacity in *Homo* began a dramatic trajectory" that resulted in an "approximate doubling in brain size."[103] Wood and Collard's review in *Science* the following year found that *only one single trait of one individual hominin fossil species qualified as "intermediate"* between *Australopithecus* and *Homo*: the brain size of *Homo erectus*.[104] However, even this one intermediate trait does not necessarily offer any evidence that *Homo* evolved from less intelligent hominids. As they explain: "Relative brain size does not group the fossil hominins in the same way as the other variables. This pattern suggests that the link between relative brain size and adaptive zone is a complex one."[105]

Likewise, others have shown that intelligence is determined largely by internal brain organization, and is far more complex than the sole variable of brain size. As one paper in the *International Journal of Primatology* writes, "brain size may be secondary to the selective advantages of allometric reorganization within the brain."[106] Thus, finding a few skulls of intermediate size does little to bolster the case that humans evolved from more primitive ancestors. (See **Figure 3-6** below.)

Similar to brain size, a study of the pelvic bones of australopithecines and *Homo* proposed "a period of very rapid evolution corresponding to the emergence of the genus *Homo*."[107] In fact, a paper in the *Journal of Molecular Biology and Evolution* found that *Homo* and *Australopithecus* differ significantly in brain size, dental function, increased cranial buttressing, expanded body height, visual, and respiratory changes and stated: "We, like many others, interpret the anatomical evidence to show that early *H. sapiens* was significantly and dramatically different from… australopithecines in virtually every element of its skeleton and every remnant of its behavior."[108]

Figure 3-6: Got a big head? Don't get a big head. Brain size not always a good indicator of intelligence or evolutionary relationships. Case in point: Neanderthals had a larger average skull size than modern humans. More- over, skull size can vary greatly within an individual species. (See Figure 3-8.) Given the range of modern human genetic variation, a progression of relatively small to very large skulls could be created by using the bones of living humans alone. This could give the misimpression of some evolution- ary lineage when in fact it is merely the intepretation of data by precon- ceived notions of what happened. The lesson is this: don't be too impressed when textbooks, news stories, or TV documentaries display skulls lined up from small sizes to larger ones.
Illustration: Jonathan Jones

Noting these many changes, the study called the origin of humans, "a real acceleration of evolutionary change from the more slowly changing pace of australopithecine evolution" and stated that such a transforma- tion would have included radical changes: "The anatomy of the earliest *H. sapiens* sample indicates significant modifications of the ancestral ge- nome and is not simply an extension of evolutionary trends in an earlier australopithecine lineage throughout the Pliocene. In fact, its combina- tion of features never appears earlier."[109]

These rapid, unique, and genetically significant changes are termed "a genetic revolution" where "no australopithecine species is obviously transitional."[110] For those not constrained by an evolutionary paradigm, what is also not obvious is that this transition took place at all. The lack of fossil evidence for this hypothesized transition is confirmed by Harvard paleoanthropologists Daniel E. Lieberman, David R. Pilbeam, and Richard W. Wrangham, who provide a stark analysis of the lack of evidence for a transition from *Australopithecus* to *Homo*:

> Of the various transitions that occurred during human evolution, the transition from *Australopithecus* to *Homo* was undoubtedly one of the most critical in its magnitude and consequences. As with many key evolutionary events, there is both good and bad news. First, the bad news is that many details of this transition are obscure because of the paucity of the fossil and archaeological records.[111]

As for the "good news," they still admit: "[A]lthough we lack many details about exactly how, when, and where the transition occurred from *Australopithecus* to *Homo*, we have sufficient data from before and after the transition to make some inferences about the overall nature of key changes that did occur."[112]

In other words, the fossil record provides ape-like australopithecines, and human-like *Homo*, but not fossils documenting a transition between them.

In the absence of fossil evidence, evolutionary claims about the transition to *Homo* are said to be mere "inferences" made by studying the non-transitional fossils we do have, and then assuming that a transition must have occurred somehow, sometime, and someplace.

Again, this does not make for a compelling evolutionary account of human origins. Ian Tattersall also acknowledges the lack of evidence for a transition to humans:

> Our biological history has been one of sporadic events rather than gradual accretions. Over the past five million years, new hominid species have regularly emerged, competed, coexisted, colonized new environments and succeeded—or failed. We have only the dimmest of perceptions of how this dramatic history of innovation and interaction unfolded…[113]

Likewise, evolutionary biologist Ernst Mayr recognized our abrupt appearance when he wrote in 2004:

> The earliest fossils of *Homo*, *Homo rudolfensis* and *Homo erectus*, are separated from *Australopithecus* by a large, unbridged gap. How can

we explain this seeming saltation? Not having any fossils that can serve as missing links, we have to fall back on the time-honored method of historical science, the construction of a historical narrative.[114]

As another commentator proposed, the evidence implies a "big bang theory" of the appearance of our genus *Homo*.[115]

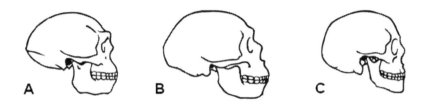

Figure 3-7: A comparison of skulls from *Homo erectus* (A), *Homo neanderthalensis* (B), and *Homo sapiens* (C).
Illustration: Adapted from Wikimedia Commons work in the public domain.

ALL IN THE FAMILY

In contrast to the australopithecines, the major members of *Homo*— such as *erectus* and the Neanderthals (*Homo neanderthalensis*)—are very similar to modern humans. (See comparison of skulls in **Figure 3-7**.) They're so similar to us that some paleoanthropologists have classified *erectus* and *neanderthalensis* as members of our own species, *Homo sapiens*.[116]

Homo erectus appears in the fossil record a little over 2 mya. The name *Homo erectus* means "upright man," and unsurprisingly, below the neck they were extremely similar to us.[117] Indeed, in contrast to the australopithecines and habilines, *Homo erectus* is the "earliest species to demonstrate the modern human semicircular canal morphology,"[118] previously noted as a feature indicative of the mode of locomotion. Another study found that total energy expenditure (TEE), a complex character related to body size, diet quality, and food-gathering activity, "increased substantially in *Homo erectus* relative to the earlier australopithecines," beginning to approach the very high TEE value of modern humans.[119]

As one paper in a 2007 Oxford University press volume notes, "despite having smaller teeth and jaws, *H. erectus* was a much bigger animal than the australopithecines, being humanlike in its stature, body mass, and body proportions."[120] While the average brain-size of *Homo erectus* is less than modern humans, *erectus* cranial capacities are well within the range of normal human variation (**Figure 3-8**).

Figure 3-8. Cranial Capacities of Extant and Extinct Hominids[121]		
Taxon	Cranial Capacities	Taxon Resembles
Gorilla (*Gorilla gorilla*)	340–752 cc	Modern Apes
Chimpanzee (*Pan troglodytes*)	275–500 cc	
Australopithecus	370–515 cc (Avg. 457 cc)	
Homo habilis	Avg. 552 cc	
Homo erectus	850–1250 cc (Avg. 1016 cc)	Modern Humans
Neanderthals	1100–1700 cc (Avg. 1450 cc)	
Homo sapiens	800–2200 cc (Avg. 1345 cc)	

Donald Johanson suggests that were *erectus* alive today, it could mate successfully with modern humans to produce fertile offspring.[122] In other words, were it not for our separation by time, we might be considered biologically as interbreeding members of the same species.[123]

Though Neanderthals have been stereotyped as bungling, primitive precursors to modern humans, in reality, they were so similar to us that if a Neanderthal walked past you on the street, you probably wouldn't notice many differences. Wood and Collard make this same point in drier, more technical language: "The numerous associated skeletons of *H. neanderthalensis* indicate that their body shape was within the range of variation seen in modern humans."[124]

Washington University paleoanthropologist Erik Trinkaus likewise argues: "They may have had heavier brows or broader noses or stockier builds, but behaviorally, socially and reproductively they were all just people."[125] In a 2007 *Washington Post* interview, Trinkaus dispelled the myth that Neanderthals were intellectually inferior:

> Although Neanderthals live in the public imagination as hulking and slow-witted "Alley Oops," Trinkaus and others say there is no reason to believe they were any less intelligent than the newly arrived 'modern humans.' Neanderthals were stockier and had larger brows,

sharper teeth and more jutting jaws, but their brain capacity appears to have been no different than that of the newcomers.[126]

But it isn't just the "public imagination" which has wrongly cast Neanderthals as unintelligent brutes. A 2003 article in *Smithsonian magazine* traces these myths back to prior European anthropologists, who, inspired by Darwin, wrongly promoted the "subhuman" view:

"In the minds of the European anthropologists who first studied them, Neanderthals were the embodiment of primitive humans, subhumans if you will," says Fred H. Smith, a physical anthropologist at Loyola University in Chicago who has been studying Neanderthal DNA. "They were believed to be scavengers who made primitive tools and were incapable of language or symbolic thought." Now, he says, researchers believe that Neanderthals "were highly intelligent, able to adapt to a wide variety of ecological zones, and capable of developing highly functional tools to help them do so. They were quite accomplished."[127]

University of Bordeaux archaeologist Francesco d'Errico affirms these comments, stating, "Neanderthals were using technology as advanced as that of contemporary anatomically modern humans and were using symbolism in much the same way."[128]

Hard evidence backs up these claims. Anthropologist Stephen Molnar explains that "the estimated mean size of [Neanderthal] cranial capacity (1,450 cc) is actually higher than the mean for modern humans (1,345 cc)."[129] One paper in *Nature* suggested, "the morphological basis for human speech capability appears to have been fully developed" in Neanderthals.[130] Indeed, Neanderthal remains have with been found associated with signs of culture including art, burial of their dead, and technology including the usage of complex tools.[131] At least one artifact shows Neanderthals made musical instruments like the flute.[132] While this example might be dated and uncertain, there is even a report in *Nature* from 1908 which claims to have discovered a Neanderthal type skeleton wearing chain mail armor.[133] Whether that report is right or wrong, it is clear Neanderthals were not intellectually dissimilar from their "human" contemporaries. As experimental archaeologist Metin

Eren said, when it came to making tools, "in many ways, Neanderthals were just as smart or just as good as us."[134] Likewise, Trinkaus says that when comparing ancient Europeans and Neanderthals: "Both groups would seem to us dirty and smelly but, cleaned up, we would understand both to be human. There's good reason to think that they did as well."[135]

One of these good reasons is the presence of "morphological mosaics—skeletons showing a mix of modern human and Neanderthal traits which indicate "that Neandertals and modern humans are members of the same species who interbred freely."[136] In 2010, *Nature* reported the finding of Neanderthal DNA markers in living humans: "A genetic analysis of nearly 2,000 people from around the world indicates that such extinct species interbred with the ancestors of modern humans twice, leaving their genes within the DNA of people today."[137] In the words of Jeffrey Long, a genetic anthropologist at the University of New Mexico, "Neanderthals didn't completely disappear" because "[t]here is a little bit of Neanderthal leftover in almost all humans."[138] Unsurprisingly, these observations have led to proposals that Neanderthals were a sub-race of our own species.[139]

We saw earlier that Leslie Aiello said "Australopithecines are like apes, and the *Homo* group are like humans."[140] This is consistent with what we see in the major members of *Homo* like *H. erectus* and Neanderthals. According to Siegrid Hartwig-Scherer, the differences between these humanlike members of the genus *Homo* can be explained as microevolutionary effects of "size variation, climatic stress, genetic drift and differential expression of [common] genes."[141] These small differences do *not* supply evidence of the evolution of humans from earlier ape-like creatures.

CONCLUSION

WHILE VIRTUALLY the entire hominin fossil record is marked by incomplete and fragmented fossils, about 3–4 mya we see ape-like australopithecines appearing suddenly. When the genus *Homo* appears around

2 mya, it also does so in an abrupt fashion, without clear evidence of a transition from previous ape-like hominins. Subsequent members of the genus *Homo* appear very similar to modern humans, and their differences amount to small-scale microevolutionary changes.

At the beginning of this chapter I quoted SMU anthropologist Ronald Wetherington telling the Texas State Board of Education that the fossil record shows an unbroken sequence documenting our gradual Darwinian evolution from ape-like species. Were we to revise Wetherington's testimony in light of the actual evidence discussed in the technical literature, we might say that the hominin fossil record is anything but unbroken. There are many gaps and virtually no plausible transitional fossils that are generally accepted, even by evolutionists, to be direct human ancestors.

Thus, public claims of evolutionists to the contrary, the appearance of humans in the fossil record appears to be been anything *but* a gradual Darwinian evolutionary process. The Darwinian belief that humans evolved from apelike species requires inferences that go beyond the evidence and is not supported by the fossil record.

ENDNOTES

1. Ronald Wetherington testimony before Texas State Board of Education (January 21, 2009). Original recording on file with author, SBOECommt-FullJan2109B5.mp3, Time Index 1:52:00-1:52:44.
2. Ibid.
3. Donald Johanson and Blake Edgar, *From Lucy to Language* (New York: Simon & Schuster, 1996), 22–23.
4. Richard Lewontin, *Human Diversity* (New York: Scientific American Library, 1995), 163.
5. Stephen Jay Gould, *The Panda's Thumb: More Reflections in Natural History* (New York: W. W. Norton & Company, 1980), 126.
6. Frans B. M. de Waal, "Apes from Venus: Bonobos and Human Social Evolution," in *Tree of Origin: What Primate Behavior Can Tell Us about Human Social Evolution*, ed. Frans B. M. de Waal (Cambridge: Harvard University Press, 2001), 68.
7. Ibid.

8. C. E. Oxnard, "The place of the australopithecines in human evolution: grounds for doubt?," *Nature*, 258 (December 4, 1975): 389–95 (internal citation removed).

9. See Alton Biggs, Kathleen Gregg, Whitney Crispen Hagins, Chris Kapicka, Linda Lundgren, Peter Rillero, National Geographic Society, *Biology: The Dynamics of Life* (New York: Glencoe, McGraw Hill, 2000), 442–43.

10. See notes 124–139 and accompanying text.

11. Sigrid Hartwig-Scherer and Robert D. Martin, "Was 'Lucy' more human than her 'child'? Observations on early hominid postcranial skeletons," *Journal of Human Evolution*, 21 (1991): 439–49.

12. For example, see Biggs *et al.*, *Biology: The Dynamics of Life*, 438; Esteban E. Sarmiento, Gary J. Sawyer, and Richard Milner, *The Last Human: A Guide to Twenty-two Species of Extinct Humans* (New Haven: Yale University Press, 2007), 75, 83, 103, 127, 137; Johanson and Edgar, *From Lucy to Language*, 82; Richard Potts and Christopher Sloan, *What Does it Mean to be Human?* (Washington D.C.: National Geographic, 2010), 32–33, 36, 66, 92; Carl Zimmer, *Smithsonian Intimate Guide to Human Origins* (Toronto: Madison Press, 2005), 44, 50.

13. Jonathan Marks, *What It Means to be 98% Chimpanzee: Apes, People, and their Genes* (University of California Press, 2003), xv.

14. Earnest Albert Hooton, *Up From The Ape*, Revised ed. (New York: McMillan, 1946), 329.

15. For a firsthand account of one paleoanthropologist's experiences with the harsh political fights of his field, see Lee R. Berger and Brett Hilton-Barber, *In the Footsteps of Eve: The Mystery of Human Origins* (Washington D.C.: Adventure Press, National Geographic, 2000).

16. Constance Holden, "The Politics of Paleoanthropology," *Science*, 213 (1981): 737–40.

17. Ibid.

18. Johanson and Edgar, *From Lucy to Language*, 32.

19. Mark Davis, "Into the Fray: The Producer's Story," PBS NOVA Online (February 2002), accessed March 12, 2012, http://www.pbs.org/wgbh/nova/neanderthals/producer.html.

20. Henry Gee, "Return to the planet of the apes," *Nature*, 412 (July 12, 2001): 131–32.

21. Phylogeny in Figure 3-1 based upon information from multiple sources, including Carl Zimmer, *Smithsonian Intimate Guide to Human Origins* (Toronto: Madison Books, 2005), 41; Meave Leakey and Alan Walker, "Early Hominid Fossils from Africa," *Scientific American* (August 25, 2003), 16; Potts and Sloan, *What Does it Mean to be Human?*, 32–33; Ann Gibbons, *The First Human: The Race to Discover our Earliest Ancestors* (New York: Doubleday, 2006); Ann Gibbons, "A New Kind of Ancestor: *Ardipithecus* Unveiled," *Science*, 326 (October 2, 2009): 36–40.

22. "Skull find sparks controversy," *BBC News* (July 12, 2002), accessed March 4, 2012, "One of Dr Senut's colleagues, Dr Martin Pickford, who was in London this week, is also reported to have told peers that he thought the new Chadian skull was from a 'proto-gorilla'."

23. Milford H. Wolpoff, Brigitte Senut, Martin Pickford, and John Hawks, "*Sahelanthropus* or '*Sahelpithecus*'?," *Nature*, 419 (October 10, 2002): 581–82.

24. Ibid.

25. Mark Collard and Bernard Wood, "How reliable are human phylogenetic hypotheses?," *Proceedings of the National Academy of Sciences (USA)*, 97 (April 25, 2000): 5003–06.

26. Ronald Wetherington testimony before Texas State Board of Education (January 21, 2009). Time Index 2:06:00-2:06:08.

27. Bernard Wood, "Hominid revelations from Chad," *Nature*, 418 (July 11, 2002):133–35.

28. Ibid.

29. Ibid.

30. Potts and Sloan, *What Does it Mean to be Human?*, 38.

31. John Noble Wilford, "Fossils May Be Earliest Human Link," *New York Times* (July 12, 2001), accessed March 4, 2012, http://www.nytimes.com/2001/07/12/world/fossils-may-be-earliest-human-link.html.

32. John Noble Wilford, "On the Trail of a Few More Ancestors," *New York Times* (April 8, 2001), accessed March 4, 2012, http://www.nytimes.com/2001/04/08/world/on-the-trail-of-a-few-more-ancestors.html.

33. Leslie C. Aiello and Mark Collard, "Our newest oldest ancestor?," *Nature*, 410 (March 29, 2001): 526–27.

34. K. Galik, B. Senut, M. Pickford, D. Gommery, J. Treil, A. J. Kuperavage, and R. B. Eckhardt, "External and Internal Morphology of the BAR 1002'00 Orrorin tugenensis Femur," *Science*, 305 (September 3, 2004): 1450–53.

35. Sarmiento, Sawyer, and Milner, *The Last Human: A Guide to Twenty-two Species of Extinct Humans*, 35.

36. Christopher Wills, *Children Of Prometheus: The Accelerating Pace Of Human Evolution* (Reading: Basic Books, 1999), 156.

37. "Fossils May Look Like Human Bones: Biological Anthropologists Question Claims for Human Ancestry," *Science Daily* (February 16, 2011), accessed March 4, 2012, http://www.sciencedaily.com/releases/2011/02/110216132034.htm.

38. Bernard Wood and Terry Harrison, "The evolutionary context of the first hominins," *Nature*, 470 (February 17, 2011): 347–52.

39. Martin Pickford, "Fast Breaking Comments," *Essential Science Indicators Special Topics* (December 2001), accessed March 4, 2012, http://www.esi-topics.com/fbp/comments/december-01-Martin-Pickford.html.

40. Aiello and Collard, "Our newest oldest ancestor?," 526–27.

41. Tim White, quoted in Ann Gibbons, "In Search of the First Hominids," *Science*, 295 (February 15, 2002): 1214–19.

42. Jennifer Viegas, "'Ardi,' Oldest Human Ancestor, Unveiled," *Discovery News* (October 1, 2009), accessed March 4, 2012, http://news.discovery.com/history/ardi-human-ancestor.html.

43. Randolph E. Schmid, "World's oldest human-linked skeleton found," MSNBC (October 1, 2009), accessed March 4, 2012, http://www.msnbc.msn.com/id/33110809/ns/technology_and_science-science/t/worlds-oldest-human-linked-skeleton-found/.

44. Ann Gibbons, "Breakthrough of the Year: *Ardipithecus ramidus*," *Science*, 326 (December 18, 2009): 1598–99.

45. Ann Gibbons, "A New Kind of Ancestor: *Ardipithecus* Unveiled," 36–40.

46. Gibbons, "In Search of the First Hominids," 1214–19.

47. Michael D. Lemonick and Andrea Dorfman, "Ardi Is a New Piece for the Evolution Puzzle," *Time* (October 1, 2009), accessed March 4, 2012, http://www.time.com/time/printout/0,8816,1927289,00.html.

48. Gibbons, "A New Kind of Ancestor: *Ardipithecus* Unveiled," 36–40. See also Gibbons, *The First Human: The Race to Discover our Earliest Ancestors*, 15 ("The excitement was tempered, however, by the condition of the skeleton. The bone was so soft and crushed that White later described it as road-kill").

49. Jamie Shreeve, "Oldest Skeleton of Human Ancestor Found," *National Geographic* (October 1, 2009), accessed March 4, 2012, http://news.nationalgeographic.com/news/2009/10/091001-oldest-human-skeleton-ardi-missing-link-chimps-ardipithecus-ramidus.html.

50. Gibbons, "A New Kind of Ancestor: *Ardipithecus* Unveiled," 36–40.

51. Esteban E. Sarmiento, "Comment on the Paleobiology and Classification of *Ardipithecus ramidus*," *Science*, 328 (May 28, 2010): 1105b.

52. Gibbons, "A New Kind of Ancestor: *Ardipithecus* Unveiled," 36–40.

53. Wood and Harrison, "The evolutionary context of the first hominins," 347–52.

54. "Fossils May Look Like Human Bones: Biological Anthropologists Question Claims for Human Ancestry."

55. John Noble Wilford, "Scientists Challenge 'Breakthrough' on Fossil Skeleton," *New York Times* (May 27, 2010), accessed March 4, 2012, http://www.nytimes.com/2010/05/28/science/28fossil.html.

56. Eben Harrell, "Ardi: The Human Ancestor Who Wasn't?," *Time* (May 27, 2010), at http://www.time.com/time/health/article/0,8599,1992115,00.html.

57. Ibid.

58. John Roach, "Fossil Find Is Missing Link in Human Evolution, Scientists Say," *National Geographic News* (April 13, 2006), accessed March 4, 2012, http://news.nationalgeographic.com/news/2006/04/0413_060413_evolution.html.

59. Seth Borenstein, "Fossil discovery fills gap in human evolution," MSNBC (April 12, 2006), accessed March 4, 2012, http://www.msnbc.msn.com/id/12286206/.

60. See Figure 4, Tim D. White, Giday WoldeGabriel, Berhane Asfaw, Stan Ambrose, Yonas Beyene, Raymond L. Bernor, Jean-Renaud Boisserie, Brian Currie, Henry Gilbert, Yohannes Haile-Selassie, William K. Hart, Leslea J. Hlusko, F. Clark Howell, Reiko T. Kono, Thomas Lehmann, Antoine Louchart, C. Owen Lovejoy, Paul R. Renne, Haruo Saegusa, Elisabeth S. Vrba, Hank Wesselman, and Gen Suwa, "Asa Issie, Aramis and the origin of *Australopithecus*," *Nature*, 440 (April 13, 2006): 883–89.

61. Ibid.

62. Ibid.

63. Borenstein, "Fossil discovery fills gap in human evolution."

64. Tim White, quoted in Gibbons, "In Search of the First Hominids," 1214–19.

65. See for example Bernard A. Wood, "Evolution of the australopithecines," in *The Cambridge Encyclopedia of Human Evolution*, eds. Steve Jones, Robert Martin, and David Pilbeam (Cambridge: Cambridge University Press, 1992), 231–40.

66. Tim White, quoted in Donald Johanson and James Shreeve, *Lucy's Child: The Discovery of a Human Ancestor* (New York: Early Man Publishing, 1989), 163.

67. Gibbons, *The First Human: The Race to Discover our Earliest Ancestors*, 86.

68. Berger and Hilton-Barber, *In the Footsteps of Eve: The Mystery of Human Origins*, 114.

69. See for example Bernard A. Wood, "Evolution of the australopithecines," 232.

70. Mark Collard and Leslie C. Aiello, "From forelimbs to two legs," *Nature*, 404 (March 23, 2000): 339–40.

71. Collard and Aiello, "From forelimbs to two legs," 339–40. See also Brian G. Richmond and David S. Strait, "Evidence that humans evolved from a knuckle-walking ancestor," *Nature*, 404 (March 23, 2000): 382–85.

72. Ibid.

73. Jeremy Cherfas, "Trees have made man upright," *New Scientist*, 97 (January 20, 1983): 172–77.

74. Richard Leakey and Roger Lewin, *Origins Reconsidered: In Search of What Makes Us Human*, (New York: Anchor Books, 1993), 195.

75. Ibid., 193–94.

76. Figure 3-7 based upon Figure 1 in John Hawks, Keith Hunley, Sang-Hee Lee, and Milford Wolpoff, "Population Bottlenecks and Pleistocene Human Evolution," *Journal of Molecular Biology and Evolution*, 17 (2000): 2–22.

77. Fred Spoor, Bernard Wood, and Frans Zonneveld, "Implications of early hominid labyrinthine morphology for evolution of human bipedal locomotion," *Nature*, 369 (June 23, 1994): 645–48.

78. See Timothy G. Bromage and M. Christopher Dean, "Re-evaluation of the age at death of immature fossil hominids," *Nature*, 317 (October 10, 1985): 525–27.

79. See Ronald J. Clarke and Phillip V. Tobias, "Sterkfontein Member 2 Foot Bones of the Oldest South African Hominid," *Science*, 269 (July 28, 1995): 521–24.

80. Peter Andrews, "Ecological Apes and Ancestors," *Nature*, 376 (August 17, 1995): 555–56.

81. Oxnard, "The place of the australopithecines in human evolution: grounds for doubt?," 389–95.

82. Yoel Rak, Avishag Ginzburg, and Eli Geffen, "Gorilla-like anatomy on *Australopithecus afarensis* mandibles suggests *Au. afarensis* link to robust australopiths," *Proceedings of the National Academy of Sciences (USA)*, 104 (April 17, 2007): 6568–72.

83. Donald C. Johanson, C. Owen Lovejoy, William H. Kimbel, Tim D. White, Steven C. Ward, Michael E. Bush, Bruce M. Latimer, and Yves Coppens, "Morphology of the Pliocene Partial Hominid Skeleton (A.L. 288-1). From the Hadar Formation, Ethiopia," *American Journal of Physical Anthropology*, 57 (1982): 403–51.

84. François Marchal, "A New Morphometric Analysis of the Hominid Pelvic Bone," *Journal of Human Evolution*, 38 (March, 2000): 347–65.

85. M. Maurice Abitbol, "Lateral view of *Australopithecus afarensis*: primitive aspects of bipedal positional behavior in the earliest hominids," *Journal of Human Evolution*, 28 (March, 1995): 211–29 (internal citations removed).

86. Leslie Aiello quoted in Leakey and Lewin, *Origins Reconsidered: In Search of What Makes Us Human*, 196. See also Bernard Wood and Mark Collard, "The Human Genus," *Science*, 284 (April 2, 1999): 65–71.

87. F. Spoor, M. G. Leakey, P. N. Gathogo, F. H. Brown, S. C. Antón, I. McDougall, C. Kiarie, F. K. Manthi, and L. N. Leakey, "Implications of new early *Homo* fossils from Ileret, east of Lake Turkana, Kenya," *Nature*, 448 (August 9, 2007): 688–91.

88. Ian Tattersall, "The Many Faces of *Homo habilis*," *Evolutionary Anthropology*, 1 (1992): 33–37.

89. Ian Tattersall and Jeffrey H. Schwartz, "Evolution of the Genus *Homo*," *Annual Review of Earth and Planetary Sciences*, 37 (2009): 67–92. Paleoanthropologists Daniel E. Lieberman, David R. Pilbeam, and Richard W. Wrangham likewise co-write that "fossils attributed to *H. habilis* are poorly associated with inadequate and fragmentary postcrania." Daniel E. Lieberman, David R. Pilbeam, and Richard W. Wrangham, "The Transition from *Australopithecus* to *Homo*," in *Transitions in Prehistory: Essays in Honor of Ofer Bar-Yosef*, eds. John J. Shea and Daniel E. Lieberman (Cambridge: Oxbow Books, 2009), 1. See also Ann Gibbons, "Who Was *Homo habilis*—And Was It Really *Homo*?," *Science*, 332 (June 17, 2011): 1370–71 ("researchers

labeled a number of diverse, fragmentary fossils from East Africa and South Africa 'H. habilis,' making the taxon a 'grab bag... a Homo waste bin,' says paleoanthropologist Chris Ruff of Johns Hopkins University in Baltimore, Maryland").

90. Alan Walker, "The Origin of the Genus Homo," in The Origin and Evolution of Humans and Humanness, ed. D. Tab Rasmussen (Boston: Jones and Bartlett, 1993), 31.

91. Ibid.

92. See Spoor et al., "Implications of new early Homo fossils from Ileret, east of Lake Turkana, Kenya," 688–91; Seth Borenstein, "Fossils paint messy picture of human origins," MSNBC (August 8, 2007), accessed March 4, 2012, http://www.msnbc.msn.com/id/20178936/ns/technology_and_science-science/t/fossils-paint-messy-picture-human-origins/.

93. Wood and Collard, "The Human Genus," 65–71.

94. Gibbons, "Who Was Homo habilis—And Was It Really Homo?," 1370–71.

95. Wood's views are described in Gibbons, "Who Was Homo habilis—And Was It Really Homo?," 1370–71. See also Wood and Collard, "The Human Genus," 65–71..

96. Spoor, Wood, and Zonneveld, "Implications of early hominid labyrinthine morphology for evolution of human bipedal locomotion," 645–48.

97. Ibid.

98. Hartwig-Scherer and Martin, "Was 'Lucy' more human than her 'child'? Observations on early hominid postcranial skeletons," 439–49.

99. Ibid.

100. Sigrid Hartwig-Scherer, "Apes or Ancestors?" in Mere Creation: Science, Faith & Intelligent Design, ed. William Dembski (Downers Grove: InterVarsity Press, 1998), 226.

101. Ibid.

102. Ibid.

103. Dean Falk, "Hominid Brain Evolution: Looks Can Be Deceiving," Science, 280 (June 12, 1998): 1714 (diagram description omitted).

104. Wood and Collard, "The Human Genus," 65–71. Specifically, Homo erectus is said to have intermediate brain size, and Homo ergaster has a Homo-like postcranial skeleton with a smaller more australopithecine-like brain size.

105. Wood and Collard, "The Human Genus," 65–71.

106. Terrance W. Deacon, "Problems of Ontogeny and Phylogeny in Brain-Size Evolution," International Journal of Primatology, 11 (1990): 237–82. See also Terrence W. Deacon, "What makes the human brain different?," Annual Review of Anthropology, 26 (1997): 337–57; Stephen Molnar, Human Variation: Races, Types, and Ethnic Groups, 5th ed. (Upper Saddle River: Prentice Hall, 2002), 189 ("The size of the brain is but one of the factors related to human intelligence").

107. Marchal, "A New Morphometric Analysis of the Hominid Pelvic Bone," 347–65.

108. Hawks, Hunley, Lee, and Wolpoff, "Population Bottlenecks and Pleistocene Human Evolution," 2–22.

109. Ibid.

110. Ibid.

111. Lieberman, Pilbeam, and Wrangham, "The Transition from *Australopithecus* to *Homo*," 1.

112. Ibid.

113. Ian Tattersall, "Once we were not alone," *Scientific American* (January, 2000): 55–62.

114. Ernst Mayr, *What Makes Biology Unique?: Considerations on the Autonomy of a Scientific Discipline* (Cambridge: Cambridge University Press, 2004), 198.

115. "New study suggests big bang theory of human evolution" University of Michigan News Service (January 10, 2000), accessed March 4, 2012, http://www.umich.edu/~newsinfo/Releases/2000/Jan00/r011000b.html.

116. See for example Eric Delson, "One skull does not a species make," *Nature*, 389 (October 2, 1997): 445–46; Hawks *et al.*, "Population Bottlenecks and Pleistocene Human Evolution," 2–22; Emilio Aguirre, "*Homo erectus* and *Homo sapiens*: One or More Species?," in *100 Years of Pithecanthropus: The Homo erectus Problem 171 Courier Forschungsinstitut Seckenberg*, ed. Jens Lorenz (Frankfurt: Courier Forschungsinstitut Senckenberg, 1994), 333–339; Milford H. Wolpoff, Alan G. Thorne, Jan Jelínek, and Zhang Yinyun, "The Case for Sinking *Homo erectus*: 100 Years of *Pithecanthropus* is Enough!," in *100 Years of Pithecanthropus: The Homo erectus Problem 171 Courier Forschungsinstitut Seckenberg*, ed. Jens Lorenz (Frankfurt: Courier Forschungsinstitut Senckenberg, 1994), 341–361.

117. See Hartwig-Scherer and Martin, "Was 'Lucy' more human than her 'child'? Observations on early hominid postcranial skeletons," 439–49.

118. Spoor, Wood, and Zonneveld, "Implications of early hominid labyrinthine morphology for evolution of human bipedal locomotion," 645–48.

119. William R. Leonard and Marcia L. Robertson, "Comparative Primate Energetics and Hominid Evolution," *American Journal of Physical Anthropology*, 102 (February, 1997): 265–81.

120. William R. Leonard, Marcia L. Robertson, and J. Josh Snodgrass, "Energetic Models of Human Nutritional Evolution," in *Evolution of the Human Diet: The Known, the Unknown, and the Unknowable*, ed. Peter S. Ungar (Oxford University Press, 2007), 344–59.

121. References for cranial capacities cited in Figure 3-11 are as follows: Gorilla: Stephen Molnar, *Human Variation: Races, Types, and Ethnic Groups*, 4th ed. (Upper Saddle River: Prentice Hall, 1998), 203. Chimpanzee: Molnar, *Human Variation: Races, Types, and Ethnic Groups*, 4th ed., 203. *Australopithecus*: Glenn C. Conroy, Gerhard W. Weber, Horst Seidler, Phillip V. Tobias,

Alex Kane, Barry Brunsden, "Endocranial Capacity in an Early Hominid Cranium from Sterkfontein, South Africa," *Science*, 280 (June 12, 1998): 1730–31; Wood and Collard, "The Human Genus," 65–71. *Homo habilis*: Wood and Collard, "The Human Genus," 65–71. *Homo erectus*: Molnar, *Human Variation: Races, Types, and Ethnic Groups*, 4th ed., 203; Wood and Collard, "The Human Genus," 65–71. Neanderthals: Molnar, *Human Variation: Races, Types, and Ethnic Groups*, 4th ed., 203; Molnar, *Human Variation: Races, Types, and Ethnic Groups*, 5th ed., 189. *Homo sapiens* (modern man): Molnar, *Human Variation: Races, Types, and Ethnic Groups*, 4th ed., 203; E. I. Odokuma, P. S. Igbigbi, F. C. Akpuaka and U. B. Esigbenu, "Craniometric patterns of three Nigerian ethnic groups," *International Journal of Medicine and Medical Sciences*, 2 (February, 2010): 34–37; Molnar, *Human Variation: Races, Types, and Ethnic Groups*, 5th ed., 189.

122. Donald C. Johanson and Maitland Edey, *Lucy: The Beginnings of Humankind* (New York: Simon & Schuster, 1981), 144.

123. Ibid.

124. See Wood and Collard, "The Human Genus," 65–71.

125. Michael D. Lemonick, "A Bit of Neanderthal in Us All?," *Time* (April 25, 1999), accessed March 5, 2012, http://www.time.com/time/magazine/article/0,9171,23543,00.html.

126. Marc Kaufman, "Modern Man, Neanderthals Seen as Kindred Spirits," *Washington Post* (April 30, 2007), accessed March 5, 2012, http://www.washingtonpost.com/wp-dyn/content/article/2007/04/29/AR2007042901101_pf.html.

127. Joe Alper, "Rethinking Neanderthals," *Smithsonian magazine* (June, 2003), accessed March 5, 2012, http://www.smithsonianmag.com/science-nature/neanderthals.html.

128. Francesco d'Errico quoted in Alper, "Rethinking Neanderthals."

129. Molnar, *Human Variation: Races, Types, and Ethnic Groups*, 5th ed., 189.

130. B. Arensburg, A. M. Tillier, B. Vandermeersch, H. Duday, L. A. Schepartz, and. Y. Rak, "A Middle Palaeolithic human hyoid bone," *Nature*, 338 (April 27, 1989): 758–60.

131. Alper, "Rethinking Neanderthals"; Kate Wong, "Who were the Neandertals?," *Scientific American* (August, 2003): 28–37; Erik Trinkaus and Pat Shipman, "Neandertals: Images of Ourselves," *Evolutionary Anthropology*, 1 (1993): 194–201; Philip G. Chase and April Nowell, "Taphonomy of a Suggested Middle Paleolithic Bone Flute from Slovenia," *Current Anthropology*, 39 (August/October 1998): 549–53; Tim Folger and Shanti Menon, "... Or Much Like Us?," *Discover Magazine*, January, 1997, accessed March 5, 2012, http://discovermagazine.com/1997/jan/ormuchlikeus1026; C. B. Stringer, "Evolution of early humans," in *The Cambridge Encyclopedia of Human Evolution*, eds. Steve Jones, Robert Martin, and David Pilbeam (Cambridge: Cambridge University Press, 1992), 248.

132. Philip G. Chase and April Nowell, "Taphonomy of a Suggested Middle Paleolithic Bone Flute from Slovenia," *Current Anthropology*, 39 (August/October 1998): 549–553; Folger and Menon, "... Or Much Like Us?"

133. Notes in *Nature*, 77 (April 23, 1908): 587.

134. Metub Eren quoted in Jessica Ruvinsky, "Cavemen: They're Just Like Us," *Discover Magazine* (January, 2009), accessed March 5, 2012, http://discovermagazine.com/2009/jan/008.

135. Erik Trinkaus, quoted in Kaufman, "Modern Man, Neanderthals Seen as Kindred Spirits."

136. Erik Trinkaus and Cidália Duarte, "The Hybrid Child from Portugal," *Scientific American* (August, 2003): 32.

137. Rex Dalton, "Neanderthals may have interbred with humans," *Nature* news (April 20, 2010), accessed March 5, 2012, http://www.nature.com/news/2010/100420/full/news.2010.194.html.

138. Ibid.

139. Delson, "One skull does not a species make," 445–46.

140. Leslie Aiello quoted in Leakey and Lewin, *Origins Reconsidered: In Search of What Makes Us Human*, 196. See also Wood and Collard, "The Human Genus," 65–71.

141. Hartwig-Scherer, "Apes or Ancestors," 220.

4

Francis Collins, Junk DNA, and Chromosomal Fusion

Casey Luskin

Leading proponents of "theistic evolution" like Francis Collins
offer two primary genetic arguments for human/ape common
ancestry: "junk" DNA and chromosomal fusion. The argument
from junk DNA fails because most non-coding DNA has
important cellular functions and is not "junk." The argument from
chromosomal fusion fails because at most it indicates that humans
experienced a fusion event, but says nothing about whether our
lineage leads back to a common ancestor with apes.

IN HIS BEST-SELLING BOOK *THE LANGUAGE OF GOD* (2006), GENETI-
cist Francis Collins claims that human DNA provides "powerful support for Darwin's theory of evolution, that is descent from a common
ancestor with natural selection operating on randomly occurring variations."[1] More specifically, he argues that our DNA demonstrates that
humans and apes share a common ancestor.

Formerly the head of the Human Genome Project, Collins is
well-known as an evangelical Christian who embraces both Darwinian evolution and embryonic stem cell research.[2] With the help of a $2
million grant from the John Templeton Foundation in 2008, Collins
co-founded the BioLogos Foundation with the purpose of persuading
Christian leaders and laypeople to accept biological evolution.[3] Collins
had to step down from the group after being appointed director of the
National Institutes of Health by President Barack Obama, but his emphatic defense of ape/human common ancestry still has wide influence
in the faith community.

Collins offers two main DNA-based arguments for his claim that humans share a common ancestor with apes and other animals. First, non-coding DNA shared by humans and other mammals is supposedly functionless junk, which according to Collins means "the conclusion of a common ancestor for humans and mice is virtually inescapable."[4] Second, human chromosome #2 resulted from the fusion of two chromosomes like those found in apes—evidence which Collins claims is "very difficult to understand… without postulating a common ancestor" between humans and apes.[5]

These are common evolutionary arguments for ape/human common ancestry, but as this chapter will show, Collins's case is based largely on outdated science and questionable assumptions. To be specific:

+ Numerous studies have found extensive evidence of function for non-coding DNA, showing that it is not genetic "junk" after all.

+ Human chromosomal fusion may imply that the human lineage experienced a fusion event, but this tells us nothing about whether our lineage extends back to share a common ancestor with apes. Moreover, the genetic evidence for human chromosomal fusion isn't nearly as strong as Collins and others make it out to be.

In sum, the evidence from DNA does not establish Collins's conclusions about human evolution.

NON-CODING DNA: NOT REALLY "JUNK" AFTER ALL?

To HIS credit, Collins avoids the usual simplistic argument that shared functional genetic similarity between two species *must* demonstrate they shared a common ancestor, acknowledging that functional genetic similarity "alone does not, of course, prove a common ancestor" because a designer could have "used successful design principles over and over again."[6] Instead, Collins offers a different argument. He cites a type of DNA called ancient repetitive elements (AREs) as allegedly non-functional "junk" DNA, which in his view demonstrates both Darwinian evolution and human/ape common ancestry.

Repetitive elements are common in mammalian genomes. We have them. Apes have them. Mice have them. And we often share them in the same places in our genomes. Collins asserts that AREs are "genetic flotsam and jetsam" which "presen[t] an overwhelming challenge to those who hold to the idea that all species were created *ex nihilo*."[7] In his view, "[u]nless one is willing to take the position that God has placed these decapitated AREs in these precise positions to confuse and mislead us, the conclusion of a common ancestor for humans and mice is virtually inescapable."[8] Sounding much like Collins, atheist Darwinist Richard Dawkins likewise writes that "creationists might spend some earnest time speculating on why the Creator should bother to litter genomes with... junk tandem repeat DNA."[9] It's worth noting that both Collins and Dawkins are making a theological argument (basically, "God wouldn't do it that way") as much as a scientific claim. I will leave the soundness of their theology to others, but their science has been overturned by the evidence.

Contrary to both Collins and Dawkins, even a cursory review of the scientific literature shows it is wildly inappropriate to simply assume that repetitive DNA—or most others types of non-coding DNA—are useless genetic "junk."

Open-minded scientists understood this long before Collins wrote his book. In 2002, biologist Richard Sternberg surveyed the literature and found extensive evidence for functions for AREs. Writing in the *Annals of the New York Academy of Sciences*, he found that ARE functions include:

- Satellite repeats forming higher-order nuclear structures
- Satellite repeats forming centromeres
- Satellite repeats and other REs involved in chromatin condensation
- Telomeric tandem repeats and LINE elements

- Subtelomeric nuclear positioning/chromatin boundary elements

- Non-TE interspersed chromatin boundary elements

- Short, interspersed nuclear elements or SINEs as nucleation centers for methylation

- SINEs as chromatin boundary/insulator elements

- SINEs involved in cell proliferation

- SINEs involved in cellular stress responses

- SINEs involved in translation (may be connected to stress response)

- SINEs involved in binding cohesin to chromosomes

- LINEs involved in DNA repair[10]

Sternberg concluded that "the selfish [junk] DNA narrative and allied frameworks must join the other 'icons' of neo-Darwinian evolutionary theory that, despite their variance with empirical evidence, nevertheless persist in the literature."[11]

Other genetic research has continued to uncover functions for various types of repetitive DNA, including SINE,[12] LINE,[13] and *Alu* elements.[14] One paper even suggested that repetitive *Alu* sequences might be involved in "the development of higher brain function" in humans.[15] Numerous other functions have been discovered for various types of non-coding DNA, including:

- Repairing DNA[16]

- Assisting in DNA replication[17]

- Regulating DNA transcription[18]

- Aiding in folding and maintenance of chromosomes[19]

- Controlling RNA editing and splicing[20]

- Helping to fight disease[21]

- Regulating embryological development[22]

Sternberg, along with University of Chicago geneticist James Shapiro, predicted in 2005 that "one day, we will think of what used to be called 'junk DNA' as a critical component of truly 'expert' cellular control regimes."[23]

The day foreseen by Sternberg and Shapiro may have come sooner than they expected. In 2007, the *Washington Post* reported that a huge scientific consortium, the ENCODE project, discovered that "the vast majority of the 3 billion 'letters' of the human genetic code are busily toiling at an array of previously invisible tasks."[24] According to an article in *Nature* reporting on the project:

> Biology's new glimpse at a universe of non-coding DNA—what used to be called 'junk' DNA—has been fascinating and befuddling. Researchers from an international collaborative project called the Encyclopedia of DNA Elements (ENCODE) showed that in a selected portion of the genome containing just a few per cent of protein-coding sequence, between 74% and 93% of DNA was transcribed into RNA. Much non-coding DNA has a regulatory role; small RNAs of different varieties seem to control gene expression at the level of both DNA and RNA transcripts in ways that are still only beginning to become clear.[25]

A 2007 paper in *Nature Reviews Genetics*, titled "Genome-wide transcription and the implications for genomic organization," explains the extensive, complex, and vital nature of these mysterious functions of non-coding DNA:

> Evidence indicates that most of both strands of the human genome might be transcribed, implying extensive overlap of transcriptional units and regulatory elements. These observations suggest that genomic architecture is not colinear, but is instead interleaved and modular, and that the same genomic sequences are multifunctional: that is, used for multiple independently regulated transcripts and as regulatory regions.[26]

Likewise, a 2008 paper in *Science* found that almost all parts of well-studied eukaryotic genomes are transcribed, yielding immense amounts of non-protein-coding strands of RNA which likely have functions:

The past few years have revealed that the genomes of all studied eukaryotes are almost entirely transcribed, generating an enormous number of non-protein-coding RNAs (ncRNAs). In parallel, it is increasingly evident that many of these RNAs have regulatory functions. Here, we highlight recent advances that illustrate the diversity of ncRNA control of genome dynamics, cell biology, and developmental programming.[27]

The paper goes on to elaborate specifically that *repetitive* elements play important roles in this cellular control: "Given the abundance of transcribed repetitive sequences, this may represent a genome-wide strategy for the control of chromatin domains that may be conserved throughout eukaryotes."[28]

A 2003 article in *Science* acknowledged that "junk DNA" labels—similar to those used by Collins—have discouraged scientists from discovering the functions of noncoding repetitive elements:

Although catchy, the term 'junk DNA' for many years repelled mainstream researchers from studying noncoding DNA. Who, except a small number of genomic clochards, would like to dig through genomic garbage? However, in science as in normal life, there are some clochards who, at the risk of being ridiculed, explore unpopular territories. Because of them, the view of junk DNA, especially repetitive elements, began to change in the early 1990s. Now, more and more biologists regard repetitive elements as a genomic treasure.[29]

Despite widespread Darwinian assumptions to the contrary, the paper concluded that, "repetitive elements are not useless junk DNA but rather are important, integral components of eukaryotic genomes."[30]

In addition to repetitive elements, another kind of "junk" DNA appealed to by Collins to support ape/human common ancestry is the "pseudogene."

Collins writes in *The Language of God* that a pseudogene in humans (caspase-12) is functionless and asks, "why would God have gone to the trouble of inserting such a nonfunctional gene in this precise location?"[31] He makes this same type of argument in his later book, *The Language of Science and Faith* (2011), citing a supposedly functionless vitamin C pseudogene in humans: "To claim that the human genome was created by God independently, rather than having descended from a common ancestor, means God inserted a broken piece of DNA into our genomes. This is not remotely plausible."[32] Similarly, Brown University biologist Kenneth Miller has cited such pseudogenes as "case-closed" evidence because "common ancestry is the only possible explanation for so many matching errors in the same gene."[33]

But are pseudogenes really functionless, broken DNA?

As with AREs, multiple functions for pseudogenes have been discovered.[34] In fact, two leading biologists writing in *Annual Review of Genetics* reported that "pseudogenes that have been suitably investigated often exhibit functional roles."[35] Likewise, a 2011 paper in the journal *RNA* titled "Pseudogenes: Pseudo-functional or key regulators in health and disease?" argues they should no longer be presumed "junk": "Pseudogenes have long been labeled as 'junk' DNA, failed copies of genes that arise during the evolution of genomes. However, recent results are challenging this moniker; indeed, some pseudogenes appear to harbor the potential to regulate their protein-coding cousins."[36]

Indeed, one study suggested that even the caspase-12 pseudogene which Collins cites[37] can produce a "CARD-only protein,"[38] a type of functional proteins in humans.[39] The study suggests that human caspase-12 interacts in some biological pathways, and encourages scientists to study the caspase-12 pseudogene to understand its function: "Since human pseudo-caspase-12 is structurally comparable to ICEBERG and COP/Pseudo-ICE [CARD-only proteins], it would be interesting to study its involvement in similar pathways."[40]

While there is much we still don't know about noncoding DNA, Collins was wrong to simply assume that the vast majority of repetitive DNA is functionless "genetic flotsam and jetsam" or that pseudogenes are "broken" DNA. A genomic revolution in the past 5–10 years has uncovered numerous functions for non-coding DNA elements. Ironically, Collins himself participated in some of this research as head of the

Human Genome Project. Perhaps that is why the year following *The Language of God* Collins started to pull back on his public promotion of the idea of junk DNA, even telling one reporter that he had "stopped using the term."[41]

Despite Collins's apparent backtracking, the BioLogos Foundation he co-founded has continued to champion the junk DNA paradigm to members of the faith community as a reason they should embrace biological evolution.[42] In reality, junk DNA is an increasingly outdated way to look at non-coding DNA, and its usefulness in proving common ancestry of humans with apes is highly suspect.

CHROMOSOMAL FUSION WITHOUT COMMON ANCESTRY

THE SECOND main argument for human/ape common ancestry made by Francis Collins is his claim that human chromosome #2 has a structure similar to what one would expect if two chimpanzee chromosomes became fused, end to end. Humans have 23 pairs of chromosomes, but chimps and other great apes have 24. In *The Language of God*, Collins argues that this chromosomal fusion explains why humans have one less pair of chromosomes than apes, claiming "it is very difficult to understand this observation without postulating a common ancestor."[43]

To the contrary, it is very easy to understand this evidence without postulating a common ancestor.

Assuming that human chromosome 2 is fused as Collins claims it is, human chromosomal fusion merely shows that at some point within our lineage, two chromosomes became fused. Logically speaking, this evidence tells us nothing about whether our human lineage leads back to a common ancestor with apes. Nor does it tell us whether the earliest humans were somehow ape-like.

Even if our ancestors did once have 24 pairs of chromosomes, they still could have been essentially just like fully modern humans. As University of North Carolina, Charlotte anthropologist Jonathan Marks observes, "the fusion isn't what gives us language, or bipedalism, or a big brain, or art, or sugarless bubble gum. It's just one of those neutral changes, lacking outward expression and neither good nor bad."[44] At best, the evidence for human chromosomal fusion implies that one of our ancestors experienced a chromosomal fusion event which then got fixed into the human population; but this evidence tells us nothing about whether we share a common ancestor with apes.

The evidence for human chromosomal fusion does not provide special evidence that humans share a common ancestor with chimps. The evidence is equally compatible with common descent (A) or common design (B) where there is no shared ancestry between the species.

If we step outside of the Darwinian box, the following scenario becomes equally possible with common ancestry:

1. The human lineage was designed separately from apes.
2. A chromosomal fusion event occurred in our lineage.
3. The trait spread throughout the human population during a genetic bottleneck (when the human population size suddenly became quite small)

In such a scenario, the evidence would appear precisely as we find it, without any common ancestry with apes, as explained by the two models described in **Figure 4-1** on the next page.

In Model A, humans and chimps share a common ancestor, and the human line experienced a chromosomal fusion event. This is the standard evolutionary model promoted by Francis Collins.

However, Model B is equally compatible with the observed data. In Model B, humans and apes do not share a common ancestor, and the human line experienced a chromosomal fusion event. This model shows that it is quite easy to explain the chromosomal fusion evidence without postulating a common ancestor without any common ancestry with apes.

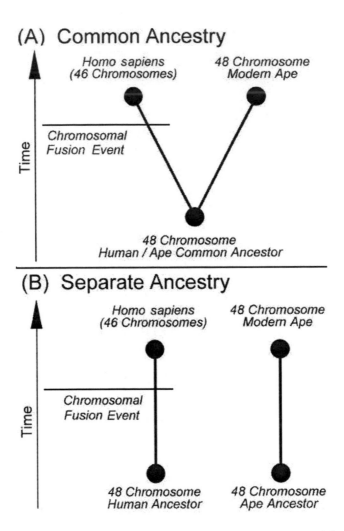

Figure 4-1: Two models for understanding human chromosomal fusion.
Illustration: Casey Luskin

To further illustrate why chromosomal fusion does not demonstrate common ancestry between humans and apes, consider the following hypothetical situation.

Imagine that in the year 2050, a small, isolated human tribe experiences a second chromosomal fusion event (they remain fertile and otherwise normal). We'll call them the "Doublefuser" people. In 2100, war, sickness, and famine destroy the rest of humanity. But the Doublefusers survive and repopulate the earth, rediscovering genetics and evolution. Eventually, the Doublefusers develop technology to examine their own chromosomes and their scientists exclaim, "We Doublefusers have 22 pairs of chromosomes, including two pairs of fused chromosomes. Since apes have 24 pairs of chromosomes, we must be descended from ape-like creatures with 48 chromosomes!"

From our vantage, we see that the Doublefusers' second chromosomal fusion event took place recently, far removed from any common ancestor between humans and chimps, and offers little logical reason to infer human-chimp common ancestry. Why should we assume the case must be any different with our one fused chromosome? Yet many Darwinian evolutionists mistakenly view our one pair of fused chromosomes precisely as the Doublefusers view their two pairs of fused chromosomes.

The Darwinian might respond by saying: "The fusion evidence shows our ancestors once had 48 chromosomes, like chimpanzees and other great apes do today. Moreover, our fused chromosome #2 even contains segments resembling ape chromosomes 2a and 2b. Common ancestry would have predicted all this evidence." But the Darwinian rejoinder merely restates the fact that humans and apes share a highly similar genetic structure. This high level of human/chimp functional genetic similarity does not demonstrate common ancestry. In Chapter 1, Ann Gauger has already elaborated why shared human/chimp functional genetic similarity does not necessarily demonstrate common ancestry: Functional genetic similarities might result from functional requirements and common design rather than inheritance from a common ancestor. Indeed, as we have seen, even Francis Collins admits that functional genetic similarity "alone does not, of course, prove a common ancestor" because a designer could have "used successful design principles over and over again."

At most, the chromosomal fusion evidence strengthens something we already knew—that chimps and humans have high genetic similarity.

Such functional similarities may just as easily be the result of functional requirements implemented via common design.

Thus far we have assumed that there really was a fusion event in human genetic history. But how strong is the actual evidence for this contention?

When using chromosomal fusion to argue for human/chimp common ancestry, biologist Kenneth R. Miller claims that "[t]he forensic case of the missing chromosome is settled beyond any doubt."[45] But in fact, the evidence for chromosomal fusion isn't nearly as clear-cut as evolutionists like Miller claim.

Telomeric DNA at the ends of our chromosomes normally consists of *thousands* of repeats of the 6-base-pair sequence TTAGGG. But the alleged fusion point in human chromosome 2 contains *far less* telomeric DNA than it should if two chromosomes were fused end-to-end: As evolutionary biologist Daniel Fairbanks admits, the location only has 158 repeats, and only "44 are perfect copies" of the sequence.[46]

Additionally, a paper in *Genome Research* found that the alleged telomeric sequences we do have are "degenerated significantly" and "highly diverged from the prototypic telomeric repeats." The paper is surprised at this finding, because the fusion event supposedly happened recently— much too recent for such dramatic divergence of sequence. Thus, the paper asks: "If the fusion occurred within the telomeric repeat arrays less than ~6 mya [million years ago], why are the arrays at the fusion site so degenerate?"[47] The conclusion is this: If two chromosomes were fused end-to-end in humans, then a huge amount of alleged telomeric DNA is missing or garbled.

Finally, the presence of telomeric DNA within a mammalian chromosome isn't highly unusual, and does not necessarily indicate some ancient point of fusion of two chromosomes. Evolutionary biologist Richard Sternberg points out that interstitial telomeric sequences (ITSs) are commonly found throughout mammalian genomes, but the telomeric sequences within human chromosome 2 are cherry-picked by evolutionists and cited as evidence for a fusion event:

[O]f all the known ITSs, and there are many in the genomes of chimps and humans, as well as mice and rats and cows... the 2q13 ITS is the only one that can be associated with an evolutionary breakpoint or fusion. The other ITSs, I hasten to add, do not square up with chromosomal breakpoints in primates. In brief, to hone in on the 2q13 ITS as being typical of what we see in the human and chimp genomes seems almost like cherry-picking data. Most are not DNA scars in the way they have been portrayed.[48]

Thus, there are at least three reasons why the evidence isn't exactly what the fusion story predicts:

- The alleged fusion point in chromosome 2 contains much less telomeric DNA than it should

- The supposed telomeric sequences we do have are highly "degenerate" and "highly diverged" from what we would expect if there were a relatively recent fusion event

- Finding interstitial telomeric DNA in mammals isn't all that remarkable, and doesn't necessarily indicate a fusion event

But—and this is the key point—even if human chromosome #2 *is* the result of two other chromosomes which became fused, this is not evidence for human/ape common ancestry. At most, it shows our human lineage experienced a chromosomal fusion event, but it does not tell us whether our lineage leads back to a common ancestor with apes.

Conclusion

IN RECENT years, genetic arguments have been offered to the public as definitive new proof that human beings share a common ancestor with apes and other animals. Francis Collins has been at the forefront of popularizing such arguments, especially in the faith community. According to Collins, there is no longer any room for disagreement: "The study of genomes leads inexorably to the conclusion that we humans share a common ancestor with other living things."[49] Indeed, not only is the idea of human/ape common ancestry beyond dispute, but "the conclusion of a common ancestor for humans and mice is virtually inescapable."[50]

Yet for all of Collins's use of terms like "inexorably" and "inescapable," the fact remains that the evidence he presents based on genetics simply does not show what he claims.

At best, the evidence discussed in this chapter reaffirms something we already knew: that humans and chimps share similar functional genetic sequences. But this can be explained by common design just as well as by common descent. What else is left? Not much.

As we have seen, Collins's arguments from junk DNA are being eroded with each passing month by new studies uncovering a myriad of functions for non-coding DNA.[51] In particular, biologists are finding extensive evidence of function for noncoding elements like ancient repetitive DNA and even pseudogenes—the precise types of DNA which Francis Collins and others claim are non-functional "genetic flotsam and jetsam" that demonstrate human/ape common ancestry

Collins's argument from chromosomal fusion also fails to deliver. Even if a chromosomal fusion event occurred, it would tell us nothing about whether our lineage leads back to a common ancestor with apes. Of course, it isn't even completely clear that a chromosomal fusion *has* occurred. Interstitial telomeric DNA doesn't necessarily indicate a fusion event, and the interstitial telomeric sequences in human chromosome 2 are "highly diverged" from what we would expect from a recent fusion event.

As a supporter of the idea that many aspects of nature are best explained by intelligent design rather than unguided processes, I want to note that intelligent design is not incompatible in principle with humans sharing ancestry with other species. At its core, intelligent design challenges not common ancestry, but the claim that life's complexity arose via unguided processes like random mutation and natural selection. Thus, a guided form of common ancestry would be compatible with intelligent design.

Nevertheless, unlike proponents of Darwinian evolution, intelligent design theorists are not obligated to accept human/ape common ancestry as a given. They are free to follow the evidence wherever it leads. And where the evidence leads is not to the conclusions promoted by Francis Collins. As we have seen, genetic arguments for human/ape common ancestry are based more upon Darwinian assumptions and outdated data than careful deductions from the evidence.

ENDNOTES

1. Francis Collins, *The Language of God: A Scientist Presents Evidence for Belief* (New York: Free Press, 2006), 127–28.
2. See: David Klinghoffer, "Francis Collins: A Biography." Wesley J. Smith, "Collins Heads NIH," *To the Point* (July 30, 2009). David Klinghoffer, "Francis Collins on Abortion," *BeliefNet* (July 8, 2009).
3. For information about the Templeton grant to launch BioLogos, see "The Language of God: BioLogos Website and Workshop," John Templeton Foundation, accessed March 19, 2012, http://www.templeton.org/what-we-fund/grants/the-language-of-god-biologos-website-and-workshop.
4. Collins, *Language of God*, 136–37.
5. Ibid., 138.
6. Ibid., 134.
7. Ibid., 136–37.
8. Ibid.
9. Richard Dawkins, "The Information Challenge," *The Skeptic*, 18 (December, 1998).
10. Richard Sternberg, "On the Roles of Repetitive DNA Elements in the Context of a Unified Genomic-Epigenetic System," *Annals of the New York Academy of Sciences*, 981 (2002): 154–88.
11. Ibid.
12. Sternberg, "On the Roles of Repetitive DNA Elements in the Context of a Unified Genomic-Epigenetic System," 154–88.
13. Tammy A. Morrish, Nicolas Gilbert, Jeremy S. Myers, Bethaney J. Vincent, Thomas D. Stamato, Guillermo E. Taccioli, Mark A. Batzer, and John V. Moran, "DNA repair mediated by endonuclease-independent LINE-1 retrotransposition," *Nature Genetics*, 31 (June, 2002): 159–65.
14. Galit Lev-Maor, Rotem Sorek, Noam Shomron, and Gil Ast, "The birth of an alternatively spliced exon: 3′ splice-site selection in Alu exons," *Science*, 300 (May 23, 2003): 1288–91; Wojciech Makalowski, "Not junk after all," *Science*, 300 (May 23, 2003): 1246–47.

15. Nurit Paz-Yaacova, Erez Y. Levanonc, Eviatar Nevod, Yaron Kinare, Alon Harmelinf, Jasmine Jacob-Hirscha, Ninette Amariglioa, Eli Eisenbergg, and Gideon Rechavi, "Adenosine-to-inosine RNA editing shapes transcriptome diversity in primates," *Proceedings of the National Academy of Sciences USA*, 107 (July 6, 2010): 12174–79.

16. Morrish *et al.*, "DNA repair mediated by endonuclease-independent LINE-1 retrotransposition," 159–65; Annie Tremblay, Maria Jasin, and Pierre Chartrand, "A Double-Strand Break in a Chromosomal LINE Element Can Be Repaired by Gene Conversion with Various Endogenous LINE Elements in Mouse Cells," *Molecular and Cellular Biology*, 20 (January, 2000): 54–60; Ulf Grawunder, Matthias Wilm, Xiantuo Wu, Peter Kulesza, Thomas E. Wilson, Matthias Mann, and Michael R. Lieber, "Activity of DNAligase IV stimulated by complex formation with XRCC4 protein in mammalian cells," *Nature*, 388 (July 31, 1997): 492–95; Thomas E. Wilson, Ulf Grawunder, and Michael R. Lieber, "Yeast DNA ligase IV mediates non-homologous DNA end joining," *Nature*, 388 (July 31, 1997): 495–98.

17. Richard Sternberg and James A. Shapiro, "How repeated retroelements format genome function," *Cytogenetic and Genome Research*, 110 (2005): 108–16.

18. Jeffrey S. Han, Suzanne T. Szak, and Jef D. Boeke, "Transcriptional disruption by the L1 retrotransposon and implications for mammalian transcriptomes," *Nature*, 429 (May 20, 2004): 268–74; Bethany A. Janowski, Kenneth E. Huffman, Jacob C. Schwartz, Rosalyn Ram, Daniel Hardy, David S. Shames, John D. Minna, and David R. Corey, "Inhibiting gene expression at transcription start sites in chromosomal DNA with antigene RNAs," *Nature Chemical Biology*, 1 (September, 2005): 216–22; J. A. Goodrich, and J. F. Kugel, "Non-coding-RNA regulators of RNA polymerase II transcription," *Nature Reviews Molecular and Cell Biology*, 7 (August, 2006): 612–16; L.C. Li, S. T. Okino, H. Zhao, H., D. Pookot, R. F. Place, S. Urakami, H.. Enokida, and R. Dahiya, "Small dsRNAs induce transcriptional activation in human cells," *Proceedings of the National Academy of Sciences USA*, 103 (November 14, 2006): 17337–42; A. Pagano, M. Castelnuovo, F. Tortelli, R. Ferrari, G. Dieci, and R. Cancedda, "New small nuclear RNA gene-like transcriptional units as sources of regulatory transcripts," *PLoS Genetics*, 3 (February, 2007): e1; L. N. van de Lagemaat, J. R. Landry, D. L. Mager, and P. Medstrand, "Transposable elements in mammals promote regulatory variation and diversification of genes with specialized functions," *Trends in Genetics*, 19 (October, 2003): 530–36; S. R. Donnelly, T. E. Hawkins, and S. E. Moss, "A Conserved nuclear element with a role in mammalian gene regulation," *Human Molecular Genetics*, 8 (1999): 1723–28; C. A. Dunn, P. Medstrand, and D. L. Mager, "An endogenous retroviral long terminal repeat is the dominant promoter for human B1,3-galactosyltransferase 5 in the colon," *Proceedings of the National Academy of Sciences USA*, 100 (October 28, 2003):12841–46; B. Burgess-Beusse, C. Farrell, M. Gaszner, M. Litt, V.

Mutskov, F. Recillas-Targa, M. Simpson, A. West, and G. Felsenfeld, "The insulation of genes from external enhancers and silencing chromatin," *Proceedings of the National Academy of Sciences USA*, 99 (December 10, 2002): 16433–37; P. Medstrand, Josette-Renée Landry, and D. L. Mager, "Long Terminal Repeats Are Used as Alternative Promoters for the Endothelin B Receptor and Apolipoprotein C-I Genes in Humans," *Journal of Biological Chemistry*, 276 (January 19, 2001): 1896–1903; L. Mariño-Ramíreza, K.C. Lewisb, D. Landsmana, and I.K. Jordan, "Transposable elements donate lineage-specific regulatory sequences to host genomes," *Cytogenetic and Genome Research*, 110 (2005):333–41.

19. S. Henikoff, K. Ahmad, and H. S. Malik "The Centromere Paradox: Stable Inheritance with Rapidly Evolving DNA," *Science*, 293 (August 10, 2001): 1098–1102; C. Bell, A. G. West, and G. Felsenfeld, "Insulators and Boundaries: Versatile Regulatory Elements in the Eukaryotic Genome," *Science*, 291 (January 19, 2001): 447–50; M.-L. Pardue & P. G. DeBaryshe, "Drosophila telomeres: two transposable elements with important roles in chromosomes," *Genetica*, 107 (1999): 189–96; S. Henikoff, "Heterochromatin function in complex genomes," *Biochimica et Biophysica Acta*, 1470 (February, 2000): O1–O8; L. M.Figueiredo, L. H. Freitas-Junior, E. Bottius, Jean-Christophe Olivo-Marin, and A. Scherf, "A central role for *Plasmodium falciparum* subtelomeric regions in spatial positioning and telomere length regulation," *The EMBO Journal*, 21 (2002): 815–24; Mary G. Schueler, Anne W. Higgins, M. Katharine Rudd, Karen Gustashaw, and Huntington F. Willard, "Genomic and Genetic Definition of a Functional Human Centromere," *Science*, 294 (October 5, 2001): 109–15.

20. Ling-Ling Chen, Joshua N. DeCerbo and Gordon G. Carmichael, "*Alu* element-mediated gene silencing," *The EMBO Journal* 27 (2008): 1694–1705; Jerzy Jurka, "Evolutionary impact of human *Alu* repetitive elements," *Current Opinion in Genetics & Development*, 14 (2004): 603–8; G. Lev-Maor *et al.* "The birth of an alternatively spliced exon: 3' splice-site selection in Alu exons," 1288–91; E. Kondo-Iida, K. Kobayashi, M. Watanabe, J. Sasaki, T. Kumagai, H. Koide, K. Saito, M. Osawa, Y. Nakamura, and T. Toda, "Novel mutations and genotype-phenotype relationships in 107 families with Fukuyama-type congenital muscular dystrophy (FCMD)," *Human Molecular Genetics*, 8 (1999): 2303–09; John S. Mattick and Igor V. Makunin, "Non-coding RNA," *Human Molecular Genetics*, 15 (2006): R17-R29.

21. M. Mura, P. Murcia, M. Caporale, T. E. Spencer, K. Nagashima, A. Rein, and M. Palmarini, "Late viral interference induced by transdominant Gag of an endogenous retrovirus," *Proceedings of the National Academy of Sciences USA*, 101 (July 27, 2004): 11117–22; M. Kandouz, A. Bier, G. D Carystinos, M. A Alaoui-Jamali, and G. Batist, "Connexin43 pseudogene is expressed in tumor cells and inhibits growth," *Oncogene*, 23 (2004): 4763–70.

22. K. A. Dunlap, M. Palmarini, M. Varela, R. C. Burghardt, K. Hayashi, J. L. Farmer, and T. E. Spencer, "Endogenous retroviruses regulate periimplantation placental growth and differentiation," *Proceedings of the National Academy of Sciences USA*, 103 (September 26, 2006): 14390–95; L. Hyslop, M. Stojkovic, L. Armstrong, T. Walter, P. Stojkovic, S. Przyborski, M. Herbert, A. Murdoch, T. Strachan, and M. Lakoa, "Downregulation of NANOG Induces Differentiation of Human Embryonic Stem Cells to Extraembryonic Lineages," *Stem Cells*, 23 (2005): 1035–43; E. Peaston, A. V. Evsikov, J. H. Graber, W. N. de Vries, A. E. Holbrook, D. Solter, and B. B. Knowles, "Retrotransposons Regulate Host Genes in Mouse Oocytes and Preimplantation Embryos," *Developmental Cell*, 7 (October, 2004): 597–606.

23. Sternberg Shapiro, "How Repeated Retroelements format genome function," 108–16.

24. Rick Weiss, "Intricate Toiling Found In Nooks of DNA Once Believed to Stand Idle," *Washington Post* (June 14, 2007), accessed March 6, 2012, http://www.washingtonpost.com/wp-dyn/content/article/2007/06/13/AR2007061302466_pf.html.

25. Erika Check Hayden, "Human Genome at Ten: Life is Complicated," *Nature*, 464 (April 1, 2010): 664–67.

26. Philipp Kapranov, Aarron T. Willingham, and Thomas R. Gingeras, "Genome-wide transcription and the implications for genomic organization," *Nature Reviews Genetics*, 8 (June, 2007): 413–23.

27. Paulo P. Amaral, Marcel E. Dinger, Tim R. Mercer, and John S. Mattick, "The Eukaryotic Genome as an RNA Machine," *Science*, 319 (March 28, 2008): 1787–89.

28. Ibid.

29. Makalowski, "Not Junk After All," 1246–47.

30. Ibid.

31. Collins, *The Language of God*, pg. 139.

32. Karl Giberson and Francis Collins, *The Language of Science and Faith: Straight Answers to Genuine Questions* (Downers Grove, IL: InterVarsity Press, 2011), 43.

33. Private correspondence with Dr. Miller.

34. See for example D. Zheng and M. B. Gerstein, "The ambiguous boundary between genes and pseudogenes: the dead rise up, or do they?," *Trends in Genetics*, 23 (May, 2007): 219–24; S. Hirotsune *et al.*, "An expressed pseudogene regulates the messenger-RNA stability of its homologous coding gene," *Nature*, 423 (May 1, 2003): 91–96; O. H. Tam *et al.*, "Pseudogene-derived small interfering RNAs regulate gene expression in mouse oocytes," *Nature*, 453 (May 22, 2008): 534–38; D. Pain *et al.*, "Multiple Retropseudogenes from Pluripotent Cell-specific Gene Expression Indicates a Potential Signature for Novel Gene Identification," *The Journal of Biological Chemistry*, 280 (Febru-

ary 25, 2005):6265–68; J. Zhang *et al.*, "NANOGP8 is a retrogene expressed in cancers," *FEBS Journal*, 273 (2006): 1723–30.

35. Evgeniy S. Balakirev and Francisco J. Ayala, "Pseudogenes, Are They 'Junk' or Functional DNA?," *Annual Review of Genetics*, 37 (2003): 123–51.

36. Ryan Charles Pink, Kate Wicks, Daniel Paul Caley, Emma Kathleen Punch, Laura Jacobs, and David Paul Francisco Carter, "Pseudogenes: Pseudo-functional or key regulators in health and disease?," *RNA*, 17 (2011): 792–98.

37. Collins acknowledges that the caspase-12 gene produces a full-fledged protein in chimps, so this is not a case where humans share a non-functional stretch of DNA with another species. In fact, the gene is not always a pseudogene in humans. According to a paper in *The American Journal of Human Genetics*, 28% of people in sub-Saharan Africa have a functioning copy of the caspase-12 gene, as do lower percentages in some other human populations. Collins ignores the obvious possibility that caspase-12 was originally designed to produce a functional protein in humans but was rendered noncoding by a mutation in some human populations at some point the recent past. See Yali Xue, Allan Daly, Bryndis Yngvadottir, Mengning Liu, Graham Coop, Yuseob Kim, Pardis Sabeti, Yuan Chen, Jim Stalker, Elizabeth Huckle, John Burton, Steven Leonard, Jane Rogers, and Chris Tyler-Smith, "Spread of an Inactive Form of Caspase-12 in Humans Is Due to Recent Positive Selection," *The American Journal of Human Genetics*, 78 (April, 2006): 659–70.

38. M. Lamkanfi, M. Kalai, and P. Vandenabeele, "Caspase-12: an overview," *Cell Death and Differentiation*, 11: (2004)365–68.

39. Sug Hyung Lee, Christian Stehlik, and John C. Reed, "COP, a Caspase Recruitment Domain-containing Protein and Inhibitor of Caspase-1 Activation Processing," *The Journal of Biological Chemistry*, 276 (September 14, 2001): 34495–500.

40. Lamkanfi, Kalai, and Vandenabeele, "Caspase-12: an overview," 365–68.

41. Collins, quoted in Catherine Shaffer, "One Scientist's Junk Is a Creationist's Treasure," *Wired Magazine Blog* (June 13, 2007), accessed March 6, 2012, .

42. See discussion in Jonathan Wells, *The Myth of Junk DNA* (Seattle: Discovery Institute Press, 2011), 98–100.

43. Collins, *The Language of God*, 138.

44. Jonathan Marks, *What it means to be 98% Chimpanzee: Apes, People, and their Genes* (Los Angeles: University of California Press, 2003), 39.

45. Kenneth R. Miller, *Only a Theory: Evolution and the Battle for America's Soul* (New York: Viking, 2008), 107.

46. Daniel Fairbanks, *Relics of Eden: The Powerful Evidence of Evolution in Human DNA* (Amherst, NY: Prometheus, 2007), 27.

47. Yuxin Fan, Elena Linardopoulou, Cynthia Friedman, Eleanor Williams, and Barbara J. Trask, "Genomic Structure and Evolution of the Ancestral Chromosome Fusion Site in 2q13-2q14.1 and Paralogous Regions on Other Human Chromosomes," *Genome Research*, 12 (2002): 1651–62.

48. Richard Sternberg, "Guy Walks Into a Bar and Thinks He's a Chimpanzee: The Unbearable Lightness of Chimp-Human Genome Similarity," *Evolution News & Views* (May 14, 2009), accessed March 6, 2012. http://www.evolutionnews.org/2009/05/guy_walks_into_a_bar_and_think020401.html (internal citations removed).

49. Collins, *The Language of God*, 133–34.

50. Ibid., 136–37.

51. For an in-depth discussion of these studies, see Wells, *The Myth of Junk DNA*.

5

THE SCIENCE OF ADAM AND EVE

Ann Gauger

Using population genetics, some scientists have argued that there
is too much genetic diversity to have passed through a bottleneck
of just two individuals. But that turns out not to be true.

IN CHAPTER 1, I ARGUED THAT OUR SIMILAR ANATOMY AND DNA
sequences are not sufficient to demonstrate that we share a common
ancestor with chimps. Using peer-reviewed scientific literature about
transitional fossils, and what is known about current chimp and human
anatomy, I concluded that there are too many anatomical changes and
too little time for neo-Darwinian processes to have accomplished the
supposed transition from our last common ancestor with chimps to us.

But the current challenge concerning our origins involves more than
fossils, anatomy, and improbable Darwinian scenarios. Now that DNA
sequencing has become relatively simple and cheap, researchers are
gathering vast amounts of human sequence data. They use the genetic
variation they find to reconstruct past events in our genetic history. They
derive evolutionary trees, estimate ancestral population sizes, and even
calculate when and where our ancestors migrated out of Africa. Based
on this kind of work, some have argued that we cannot have come from
just two first parents.

This argument directly contradicts the traditional belief of many
Christians that humanity started with an original couple, Adam and
Eve. Those affiliated with groups like the BioLogos Foundation have
gone so far as to say that Christians must abandon a belief in Adam and
Eve as sole parents of the human race, because scientific arguments sup-
posedly have disproven the possibility of their existence.

Now, I am a scientist, and not a theologian, but I feel obligated to speak. The challenge being posed to two first parents is a scientific one, so it deserves a scientific response. My purpose in this chapter is not to engage in Biblical interpretation or to pass judgment on the various views Christians hold about Adam and Eve. Instead, I propose to focus on the scientific argument and its validity.

Population genetics arguments against Adam and Eve come in many forms. Here I will examine one of the strongest cases against a first couple—the argument based on genetic variation in human leukocyte antigen (HLA) genes, some of the most variable genes in the human genome. When I began this study, I was prepared to accept that there was too much genetic diversity among these genes to have passed through just two first parents. To my surprise, I found that even this most polymorphic (most varied) region of our genome does not rule out the possibility of a first couple. And even more, buried within this region is evidence that suggests something more than common descent is responsible for our genetic make-up.

The science here is complex. In order to critically assess the arguments being made, I have had to include a fair amount of technical discussion. I realize that parts of the chapter may be challenging to some readers, but I try to provide a clear statement of my major points in non-technical language along the way.

HLA GENES

HLA GENES are involved in immune defense—they bind and present foreign peptides on the surface of immune cells (leukocytes), in order to trigger a response by other immune cells. A number of these HLA genes are present in mammals, presumably to provide immunity against a wide variety of diseases and parasites. **Figure 5-1 shows the location of the main HLA genes in humans.**

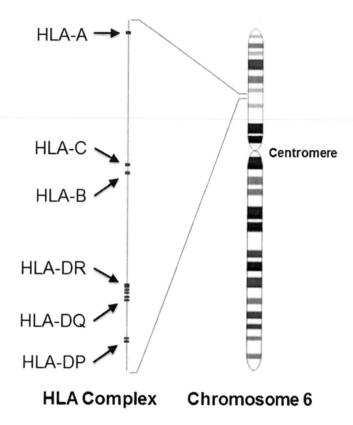

HLA Complex   Chromosome 6

Figure 5-1: The human HLA genes.
Illustration: Ann Gauger (redrawn from a Wikipedia Commons illustration, public domain)

There are many versions (alleles) currently known for each HLA gene. Because of this, the HLA complex represents one of the most difficult challenges to the idea that we came from just two first parents. If there are literally hundreds of alleles for these genes in the present human population, where did they come from? Two people can pass on at most four versions. Did all those alleles come from just two individuals with four or fewer ancestral versions?

To answer that question, I need to explain something about the methods being used in these studies, and what their underlying assumptions are.

What Is Population Genetics?

In the 1930s and 40s, Darwin's theory of evolution and Mendel's theory of genetics were combined, creating what is now called the Modern Synthesis, or what I prefer to call neo-Darwinism. Instead of focusing on how different animal forms might have evolved over time, neo-Darwinists began focusing on how genetic variation spread through **populations.** These "population geneticists," as they were called, developed mathematical models to extrapolate from existing genetic variation in populations to what may have happened to those populations in the past.

Because all these models have their roots in Darwinism, they assume that natural selection acting on **stochastic processes** (processes that occur at random, and without consideration for the organism's needs) is sufficient to explain all evolutionary change. The stochastic processes that generate genetic variation include **mutation** (changes to the DNA sequence), and **recombination** (rearrangement or swapping of genetic information between chromosomes). **Genetic drift** (the stochastic loss of genetic information due to failure to reproduce) tends to reduce the power of natural selection to drive change, especially in populations of a million or less.

Note that for neo-Darwinism, there is no room for direction or guidance in evolution. Random genetic variation occurs by chance, without any provision for the organism's needs. Natural selection does the winnowing, and genetic drift throws in a dash of additional randomness as to which variants actually survive and spread through the population.

The equations of population genetics require certain simplifications in order to make the mathematics work. Most models that use current genetic diversity to retroactively model past events assume a constant background mutation rate, with no strong selection biasing genetic change. They assume a constant population size with no migration in or out, and they assume that common descent is the underlying cause of sequence similarity. All these assumptions are subject to question, as we shall see.

The Population Genetics
Challenge to Two First Parents

In the 1990s a population biologist named Francisco Ayala set out to challenge the idea of two individual first parents, using sequence information from one of the HLA genes.[1] Ayala chose HLA-DRB1 to make his point, because at that time there were already hundreds of different versions of HLA-DRB1 known. He had reason to suspect, therefore, that there might have been considerable diversity in HLA-DRB1 at the time chimp and human lineages supposedly diverged.

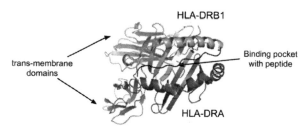

Figure 5-2: Peptide-binding domain of HLA-DR.
Illustration: Ann Gauger, created using MacPymol and PDB 1aqd

What does HLA-DRB1 do, and why is it so variable? The HLA-DRB1 protein combines with another protein called HLA-DRA to form a dimeric protein (seen above in **Figure 5-2**) called HLA-DR. (A dimer is a protein composed of two subunit proteins.) This protein dimer is embedded in the cell membrane of antigen presenting cells (a certain type of cell in the immune system). The dimer forms a peptide-binding pocket that binds foreign peptides, and presents them to other immune cells in order to trigger the production of appropriate antibodies.

The reason why there are so many variants of HLA-DRB1 is that lots of variation in the peptide-binding pocket ensures that many different foreign peptides can be recognized and bound. This is a good thing because it strengthens immunity. If a new parasite or disease-causing microbe comes along, the chances are increased that some individual will have an allele of HLA-DRB1 able to bind the invaders' broken-up proteins, and trigger the immune system to mount a defense against them.

Here's the interesting thing. Nearly all the genetic variation in the HLA-DR dimer, and thus the variation in peptides that can be bound, comes specifically from just one portion of the HLA-DRB1 gene, namely exon 2.[2] The rest of HLA-DRB1 or the HLA-DRA gene do not vary much.

Ayala obtained chimp, human and macaque DNA sequences from just exon 2 of HLA-DRB1, and reconstructed the phylogenetic history of those sequences using population genetics algorithms.[3] He drew an evolutionary tree that most closely fit the pattern of genetic variation in exon 2. Then using estimates from other sources for the average mutation rate, and the time that chimps and humans last shared a common ancestor, he calculated how far back on his tree that point of common ancestry was. Drawing a line across the tree at the point, he counted how many ancestral branches he crossed. That gave him an retrospective estimate of how many HLA-DRB1 alleles there must have been in the population at the time of the chimp/human last common ancestor.[4]

To illustrate the basic process he followed, I have drawn a simple example of a phylogenetic gene tree (**Figure 5-3**). To the left is the oldest part of the tree. As time passes, the single gene duplicates and diverges, then splits again several more times. The final number of duplicates on the right is five (A-E).

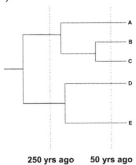

250 yrs ago   50 yrs ago

Figure 5-3: Using phylogenetic trees to estimate lineages.

Illustration: Ann Gauger

Normally population geneticists make the length of each horizontal line proportional to the amount of genetic change. The longer the length, the more nucleotide differences there are. Assuming the nucleotide differences are due to mutation over time, and assuming that mutations occur at a constant rate (not a sure thing, by the way), one can count backward to an estimated time in evolutionary history (in this case 50 and 250 years ago), and draw a line vertically through the tree. The number of lineages crossed by the line determine how many separate lineages were present at each particular time. If all estimates are correct, for this tree there would have been five lineages 50 years ago, and two lineages 250 years ago.

Following this procedure, Ayala calculated that there were *thirty-two separate versions* of the entire HLA-DRB1 gene present at the estimated time of our last common ancestor with chimps four to six million years ago (also not a sure thing—these estimates keep changing). In order for all those variant alleles to make it to modern times, he further estimated that the minimum size of the ancestral population was no fewer than 4,000, with a long-term average effective population size of 100,000.[5] This large number is necessary in a steady state population model like Ayala's. Under such conditions, assuming random mating and genetic drift, alleles are likely to be lost over time, so a large starting population is necessary to guarantee continued transmission of all the alleles. Because of this minimal estimate of 4,000, Ayala claimed that at no time was it possible for the human population to have passed through a bottleneck of two. In his view, there is just too much ancestral diversity in HLA-DRB1.

THE CHALLENGE TO THE CHALLENGE

LET'S STEP back and examine how Ayala's analysis was done. His claims against a literal Adam and Eve are based on population genetics models for how gene frequencies change in populations over time, and how ancestral gene lineages tend to coalesce. The equations used to reconstruct these trees, and to calculate ancestral population sizes, depend on simplifications and assumptions to make the mathematics tractable, as I said before. These explicit assumptions include a **constant background mutation rate** over time, **lack of selection** for genetic change on the DNA sequences being studied, **random breeding** among individuals, **no migrations** in or out of the breeding population, and a **constant population size.** If any of these assumptions turn out to be unrealistic, the results of a model may be seriously flawed.

There are also *hidden* assumptions buried in population genetics models, assumptions that rely upon the very thing they are meant to demonstrate. For example, tree-drawing algorithms *assume* that a tree of common descent exists. The population genetics equations also *assume* that random processes are the only causes of genetic change over time, an assumption drawn from naturalism. What if non-natural causes, or even unknown natural causes that do not act randomly, have intervened to produce genetic change?

It turns out that the particular DNA sequence from HLA-DRB1 that Ayala used in his analysis was *guaranteed* to give an overestimate, because he inadequately controlled for two of the above assumptions— the assumption that there is a lack of selection for genetic change on the DNA sequence being studied, and the assumption of a constant background mutation rate over time. HLA-DRB1 is known to be under strong selection for heterozygosity, meaning that having two different versions of the gene gives you a better chance of dealing with parasites and disease. Not only that, the particular region of the gene Ayala studied (exon 2) appears to have a mutation rate much higher than the background mutation rate. In fact, it is the most variable region of one of the most variable genes in our genome, and it may be a **hotspot for gene conversion** (a kind of mutation particularly likely to confuse assumptions of common descent and parsimony in tree-drawing), as we will see. Ayala did use a mathematical fudge factor for the first problem, but did not correct for the second problem.

A later study by Bergström *et al.*[6] examined the same HLA-DRB1 gene, but used intron 2, a portion of the gene not translated into protein. They chose the intron next to exon 2 expressly to avoid the confounding effects of strong selection, a high mutation rate, and/or gene conversion. They verified that this intron had a mutation rate close to the genomic background. In contrast to Ayala's study, this study concluded that only *seven* versions of the gene existed in the ancestral population from which both chimps and humans supposedly came around 4–6 million years ago, and that the population had an estimated size of 10,000 rather than the 100,000 estimated by Ayala.

In other words, by being careful about just two of the above assumptions, these researchers arrived at a dramatically lower estimate for the number of HLA-DRB1 alleles in the ancestral population than the number Ayala found in his study (i.e. seven alleles *versus* thirty-two). But the problems with Ayala's model go even deeper, as we shall see in the next section.

Phylogenetic confusion

Ayala created his phylogenetic tree based on *exon 2* sequences of the HLA-DRB1 genes, while Bergström *et al.* used *intron 2* sequences. A third study by Doxiadis *et al.* examined the phylogenetic histories of chimp, macaque and human HLA-DRB1 genes again, but this time using sequences taken either from exon 2 or introns 1-4. Surprisingly, the tree alignments using exon 2 or using introns 1-4 give markedly different pictures of the gene's phylogenetic history, even though both sets of sequences come from the very same genes. There is a substantial difference in the phylogenetic relationships. Exon 2 comparisons typically showed cross-species associations, while intron comparisons showed within-species associations.[7]

A simplified illustration of the discordant phylogenetic trees is shown in **Figure5-4** below. (For the actual trees, see Doxiadis *et al.*[8])

It is clear that the intron sequences group according to species, whereas exon 2 sequences show no species-dependent relationships.

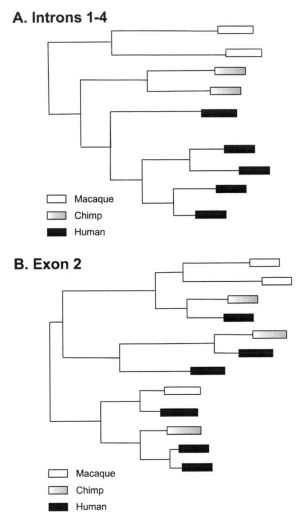

Figure 5-4: Discordant intron- and exon-based phylogenetic trees.
Illustration: Ann Gauger

This should be surprising. Although trees based on gene comparisons sometimes do not show the same phylogenetic relationships as the species themselves do, as is the case for the exon 2 sequences, when this happens it indicates something unusual is going on.

It's even more unusual that trees drawn from adjacent *segments* of the same gene disagree with one another. It's not that exon 2 is highly variable and the introns are more conserved, because this is not the case. Intron lineages can differ quite a bit from one another. **Rather, the intron lineages group together according to species, while the exon 2 lineages do not.**

Some evolutionary biologists try to explain this discordance between the HLA-DRB1 trees by arguing that this proves that these genes have their origin in deep time, before the lineages of chimps, humans and macaques separated, and that it is the exon 2 data that defines the gene's history.[9] Others think that there has been cross-species shuffling of ancient peptide-binding motifs between different exon 2 sequences over time, but leaving the intron lineages unchanged.[10] It is not clear, however, how such a patchwork cross-species assortment of exon 2 sequences could have been acquired without disrupting the species-specific introns. Furthermore, this would require that the incipient species' populations intermingled for a prolonged period of time. The intermingling is highly unlikely to have lasted for thirty million years, though, which is the last time macaques, chimps, and humans supposedly shared a common ancestor. And the fact that the intron sequences do associate by species, with branch lengths as long or longer than the exon branch lengths, argues that many of these intronic lineages have been evolving independently for quite a while, indeed some as long as thirty to forty million years. Therefore this phylogenetic discordance is something that cannot be explained by common ancestry, especially when one considers an additional piece of information: The HLA-DRB1 region of chromosome six shows little or no signs of recombination.

SPECIAL BEHAVIOR, SPECIAL DESIGN?

HLA-DRB1's CLOSEST neighbors, HLA-DQB and HLA-DQA, also bind and present foreign peptides to other immune cells, like HLA-DRB1. According to Raymond *et al.*, this region shows extreme linkage disequilibrium, meaning that there is little or no reciprocal recombination between these genes.[11]

This lack of recombination is highly unusual because it extends over 80,000 bases of DNA.[12] Stretches of DNA that do not undergo genetic recombination are called *haplotypes*. Normally, given the supposed age of these haplotypes, recombination should have occurred roughly every 150 nucleotides. Recombination does take place elsewhere in the region, just not in the vicinity of HLA-DRB1.

Despite the fact that there are hundreds of alleles for each HLA gene, only certain combinations of alleles of HLA-DQ and HLA-DR tend to occur together— they are inherited *as a block*. It may be that particular combinations of alleles work especially well together, while other less favorable combinations are removed from the population by natural selection. Alternatively, recombination may be suppressed by some other mechanism.

These co-inherited combinations of alleles constitute the basic haplotypes of HLA-DRB1. Most researchers now agree that there are *just five* of these basic haplotypes in humans. Which HLA-DRB1 gene a particular haplotype has tends to specify the particular allelic combinations of other genes in the haplotype. Based on the amount of background genetic change in the introns, three haplotypes appear to be ancient, going back thirty million years or more. These are the haplotypes we have in common with chimps and macaques. Two haplotypes are more recent, based on their accumulated background mutation, and date back to about four to six million years ago.[13] Thus, depending on when one places the time of the proposed divergence, there may have been *as few as three ancestral haplotypes, or as many as five*, when hominins diverged.

Take Home Message

Here is the whole point in simple language. The argument from population genetics has been that there is too much genetic diversity to pass through a bottleneck of two individuals, as would be the case for Adam and Eve. But that turns out not to be true.

In fact, when all the data are considered, there are just five basic versions of the HLA haplotype. *Three appear to be ancient*, pre-dating any supposed evolutionary split between chimps and humans, and *two are more recent* (some time before or after the putative most recent common ancestor of primates and humans, depending on where you draw the line). At least one of these five haplotypes appears to be missing in chimps. Given the difficulties involved in estimating the times of divergence due to the unusual genetic behavior of the region, it is possible that four or fewer of those haplotypes pre-date our supposed divergence from chimps.

Each person carries *two copies* of the Class II haplotype, so each person can carry two different alleles of HLA-DRB1. Therefore, those four haplotypes could potentially be carried by just two individuals. **This means that a first couple could have carried sufficient genetic diversity to account for four basic haplotypes**, especially given the possibility of rapid population expansion afterward.

We have dropped from an estimated 32 lineages based on DRB1 exon 2 comparisons, to seven lineages using DRB1 intron 2 comparisons, and then to between three and five ancestral haplotypes, when the whole region is considered. This is a remarkable reversal. What once seemed to be a rock-solid argument against the existence of a first couple has now dwindled considerably. The genetic analysis indicates that a first couple is possible. At the very least it is fair to say that HLA haplotype diversity cannot rule out two first parents.

What about the problem of genetic drift, and the concomitant need for a large population to prevent loss of variant haplotypes? That problem applies for a steady state, constant-size population model, but not in the case where rapid population growth is taking place. In the case of a newly emerging (created) species, rapid expansion would make it possible for all haplotypes to be preserved. In fact, there is evidence that HLA diversity increases rapidly after a new population is founded, though not usually to this degree.[14]

Now I would like to move in a more challenging direction. What if our sequence similarities are *not* the result of common descent? What if we began from two intelligently designed first parents? Is there any evidence in the data I have presented to indicate that this might be the case? If so, all this analysis of how many ancient haplotypes we share with chimps doesn't really matter.

There certainly are surprising patterns of genetic variation within HLA-DRB1 that suggest unknown processes may be operating. Let me propose that a process exists which generates specific hypervariability within exon 2 and suppresses recombination elsewhere. The process is targeted to generate diversity precisely in the peptide-binding domain. I suggest that intelligent design had to be involved at the beginning, in order to rapidly generate HLA diversity after the foundation of our new species (assuming we came from two first parents). Evidence supporting this idea comes from the fact that HLA-DRB1 diversity has in fact increased very rapidly by anyone's count, going from a handful of variants to over six hundred alleles in six million years or less. Also, the HLA-DRB1 variable regions in exon 2 show a patchwork, cross-species relationship to their surrounding DNA sequences, making their origin hard to account for by common descent. Their repeated use of similar motifs from different species may instead indicate common design. I further suggest that this process may be human-specific, since other primates don't show nearly the same degree of allelic diversity within lineages as humans do.[15]

This proposal can be supported at least in part by published data. Both gene conversion and hypermutation are known to generate antibody diversity in other immune cell lineages.[16] Sequence analysis of HLA-DRB1 alleles reveal that "recombination events either strictly located at exon 2 or involving adjacent introns have occurred" and "indicate that interlineage recombinations may be hidden and are perhaps more frequent than currently expected."[17] Others have identified sequence features thought to be involved in recombination processes, some of which are highly conserved across HLA-DRB1 alleles.[18]

In addition, several human population studies indicate that many HLA Class I and Class II genes undergo rapid interallelic recombination. For example Hedrick and Kim report that:

> new alleles that appear to be the result of microrecombination between other alleles have been found in South American Amerindians and other populations. Because the Americas have probably been populated for only the last 10,000 to 20,000 years (~ 1000 human generations), the new variants, which do not appear in Asian samples, must have arisen in this period. [19]

These include several novel variants in HLA-DRB1, HLA-DPB1, and HLA-B.[20] Hedrick and Kim go on to say:

> there is direct evidence that the rate of microrecombination at some MHC loci is high…. Zangenberg et al. (1995) examined the rate of interallelic gene conversion at the HLA-DPB1 locus in sperm from males heterozygous for six regions of the highly variable exon 2. In 111,675 sperm, they observed nine interallelic conversions for a rate of 0.81×10^{-3}, nearly 1 in 10,000 gametes.[21]

Given this data, it seems not unreasonable to propose that HLA-DRB1 diversity is the result of a process that generates specific hypervariability and/or gene conversion within exon 2 in order to rapidly generate HLA diversity. The existence of such a process essentially demolishes any population genetics arguments about ancestral population sizes.

The HLA story illustrates well the strengths and the limitations of science. Scientific claims are provisional, always subject to revision. In particular, retrospective calculations should be treated with caution, because of the number of unknown variables and hidden assumptions involved. Where ancient genetic history is involved, dogmatic statements are out of place. We understand very little of our own genetic makeup—way too little to make accurate calculations about our distant genetic past. But there are still plenty of interesting things to discover, and new proposals to consider.

Reconsidering the Evolutionary Story

I chose to look at the HLA-DRB1 story because it seemed to provide the strongest case from population genetics against two first parents. If it were true that we share thirty-two separate lineages of HLA-DRB1 with chimps, it would indeed cause difficulties for an original couple. But as we have seen, the data indicate that it is possible for us to have come from just two first parents.

Moreover, the data indicate that *DNA similarity is not going to be a simple story to unravel.* There are already regions of human DNA known to more closely resemble gorilla sequences than chimp sequences.[22] Now we have sequences that resemble macaque DNA, a primate not part of the hominid group. Furthermore, when adjacent regions of DNA yield different evolutionary trees, linked to species that diverged well before the putative most recent common ancestor of chimps and humans, something unusual is going on.

This result was a surprise to me, and threw me back into a consideration of the *whole story* of our common descent from ape-like ancestors. I already knew from my own research that similarity of form or structure was not enough to demonstrate that neo-Darwinian common descent was possible. I knew that genuine protein innovations were beyond the reach of naturalistic processes. I therefore began to re-examine everything I knew or thought I knew about human origins. I reviewed paleoanthropology, evolutionary psychology and population genetics research articles, I reviewed popular books and textbooks. I applied strict logic to the story of what would be required for our evolution from great apes. As a result of all this reading and reflection, although I was always skeptical about the plausibility of human evolution by neo-Darwinian means, I have now come to wonder about the extent of common descent as well.

Currently, neo-Darwinism is the accepted explanation for our origin. It may be, though, that as we continue to investigate our own genomes, the Darwinian explanation for our similarity with chimps—namely, common descent—will evaporate. We may discover additional features in our genome that defy explanation based on common ancestry. As evidence of common descent's insufficiency as a theory grows, alternate theories will need to be tested.

But one thing is clear right now: **Adam and Eve have not been disproven by science, and those who claim otherwise are misrepresenting the scientific evidence.**

ENDNOTES

1. Ayala was not the only one to do this. See N. Takahata , "Allelic Genealogy and human evolution," *Mol Biol Evol* 10 (1993): 2–22.
2. Briefly, HLA-DRB1 has six exons (the coding regions) interspersed by noncoding DNA, called introns.
3. Phylogenetics is the study of evolutionary relationships among organisms. These relationships are often represented as branching trees. Starting with the *assumption that common descent is true*, scientists compare the distribution of varying anatomical traits or DNA sequences that they are studying. Using mathematical algorithms, they look for tree-branching patterns that minimize conflict, or represent the fewest changes over time, but that can explain the observed distribution of traits or DNA variation.
4. Francisco Ayala, "The myth of Eve: Molecular biology and human origins," *Science* 270 (1995): 1930–1936.
5. H. A. Erlich *et al.*, "HLA sequence polymorphism and the origin of humans," *Science* 274 (1996): 1552–1554.
6. T. F. Bergström *et al.*, "Recent origin of HLA-DRB1 alleles and implications for human evolution," *Nature Genetics* 18 (1998): 237–242.
7. G. Doxiadis et al., "Reshuffling of ancient peptide binding motifs between HLA-DRB multigene family members: Old wine served in new skins," *Molecular Immunology* 45 (2008): 2743–2751.
8. Ibid.
9. J. Klein, A. Sato, and N. Nikolaidis, "MHC, TSP, and the Origin of Species: From Immunogenetics to Evolutionary Genetics," *Annu. Rev. Genet.* 41 (2007): 281–304.
10. Doxiadis, "Reshuffling of ancient peptide binding motifs."
11. C.K. Raymond *et al.*, "Ancient haplotypes of the HLA Class II region," *Genome Research* 15 (2005): 1250–1257.
12. There is an illustration of HLA-DRB1 and its neighboring genes in C. K. Raymond *et al.*, "Ancient haplotypes," 1251.
13. G. Andersson, "Evolution of the human HLA-DR region," *Frontiers in Bioscience* 3 (1998): d739–745.
14. V. Vincek, et al., "How Large Was the Founding Population of Darwin's Finches?" *Proc. R. Soc. London Ser. B* 264 (1997): 111–118.

15. G. Doxiadis et al., "Extensive DRB region diversity in cynomolgus macaques: recombination as a driving force," *Immunogenetics* 62 (2010): 137–147.
16. Ziqiang Li, Caroline J. Woo, Maria D. Iglesias-Ussel, et al., "The generation of antibody diversity through hypermutation and class switch recombination," *Genes Dev.* 18 (2004): 1–11.
17. Katja Kotsch and Rainer Blasczy, "Interlineage Recombinations as a Mechanism of The Noncoding Regions of HLA-DRB Uncover HLA Diversification," *J Immunol* 165 (2000): 5664–5670.
18. Jenny von Salomé and Jyrki P Kukkonen, "Sequence features of HLA-DRB1 locus define putative basis for gene conversion and point mutations," *BMC Genomics* 9 (2008): 228, accessed March 6, 2012, doi:10.1186/1471-2164-9-228.
19. P. W. Hedrick and T. Kim, "Genetics of Complex Polymorphisms: Parasites and Maintenance of the Major Histocompability Complex Variation," in R. S. Singh and C. B. Crimbas, editors, *Evolutionary Genetics: from Molecules to Morphology* (New York: Cambridge University Press, 2000), 211–212.
20. E. A. Titus-Trachtenberg, et al., "Analysis of HLA Class 11 Haplotypes in the Cayapa Indians of Ecuador: A Novel DRB1 Allele Reveals Evidence for Convergent Evolution and Balancing Selection at Position 86," *Am. J. Hum. Genet.* 55 (1994):160–167.
21. Hedrick and Kim, "Genetics of Complex Polymorphisms"; Gabriele Zangenberg, et al., "New HLA–DPB1 alleles generated by interallelic gene conversion detected by analysis of sperm," *Nature Genetics* 10 (1995): 407–414, accessed March 6, 2012, doi:10.1038/ng0895-407.
22. A. Hobolth, O. F. Christensen, T. Mailund, M. H. Schierup, "Genomic Relationships and Speciation Times of Human, Chimpanzee, and Gorilla Inferred from a Coalescent Hidden Markov Model," *PLoS Genet* 3 (2007): e7, accessed March 6, 2012, doi:10.1371/journal.pgen.0030007.

AUTHORS

DOUGLAS AXE

DOUGLAS AXE is Director of Biologic Institute, a research organization that develops and tests the scientific case for intelligent design in biology and explores its scientific implications. Dr. Axe's research uses both experiments and computer simulations to examine the functional and structural constraints on the evolution of proteins and protein systems. After receiving his Ph.D. from Caltech, Dr. Axe held postdoctoral and research scientist positions at the University of Cambridge, the Cambridge Medical Research Council Centre, and the Babraham Institute in Cambridge. His work has been reviewed in *Nature* and published in several peer-reviewed scientific journals, including the *Proceedings of the National Academy of Sciences*, the *Journal of Molecular Biology*, *BIO-Complexity*, *PLoS ONE*, and *Biochemistry*.

ANN GAUGER

ANN GAUGER is Senior Research Scientist at Biologic Institute. Her work uses molecular genetics and genomic engineering to study the origin, organization and operation of metabolic pathways. Dr. Gauger received a B.S. in biology from MIT and a Ph.D. in developmental biology from the University of Washington, where she studied Drosophila embryogenesis. As a post-doctoral fellow at Harvard University, she cloned and characterized the Drosophila kinesin light chain. Her research has been published in such peer-reviewed journals as *Nature*, *Development*, the *Journal of Biological Chemistry*, and *BIO-Complexity*.

Casey Luskin

Casey Luskin is Research Coordinator at Discovery Institute's Center for Science and Culture. He holds graduate degrees in both science and law. He earned his B.S. and M.S. in Earth Sciences from the University of California, San Diego. His law degree is from the University of San Diego. He formerly conducted geological research at the Scripps Institution for Oceanography (1997–2002). He has published in both science and law journals, and he has been interviewed on the evolution debate for *Nature and Science* as well as the *New York Times*, NPR, *USA Today*, and FoxNews.